ENLIGHTENMENT HUMANISM

ENLIGHTENMENT HUMANISM

ITS ADVOCATES SPEAK

SERGE BOURGAINE

ORBIS
UNUM
PRESS

Library of Congress Control Number: 2020908341

ISBN 978-0-9994906-3-1 (ebook)
ISBN 978-0-9994906-6-2 (print book)

βopγηv05ρα

With ardent appreciation to the individuals whose expressions of Enlightenment principles appear within these pages.

CONTENTS

PROLOGUE

Collections of quotations warrant the briefest of introductions. For, in all instances, it is the acumen of the contributors that merits consideration, not the prefatory chitchat of the editor. With this thought in mind, it suffices to point out that this cento serves a single aim. It celebrates Enlightenment humanism, a philosophy that draws its principles from the eponymous revolution that burst forth in the seventeenth century. Readers acquainted with the philosophy will miss little if they proceed directly to the quotations, which begin in the next chapter. For those less familiar, a synopsis follows. It begins with a definition.

Enlightenment humanism is a philosophy of life that accords to our species the faculty, guided by no supernatural authority, to lead happy, ethical, and meaningful lives.

Pause a moment to consider the uniqueness of this statement. Few such declarations, if any, gained a foothold in pre-Enlightenment societies. In fact, the opposite prevailed. Civilization after civilization relied on the timeworn practice of fabricating deities who, according to their framers, regulated the world and all life upon it from supernatural will. In this manner, gods beyond number came into existence, each offering to their inventors the façade of explanation, protection, or comfort.

Accompanying the gods came legions of self-appointed, earthly apostles, each ready to parrot bromidic answers to life's assorted enigmas. Why were the rains plentiful one year, yet absent the next? Why did one member of the community fall ill, but not another? How came the earth into being? Why do we exist? How can one differentiate good from evil? In all cases, mystical interpretations, spurious explanations, and trite *Goddidits* dominated the responses.

Centuries of such indoctrination led to an all-too-predictable result. Few could conceive a world without a creator. Few could contemplate a morality absent a deity. And few could envision a life free from theistic authority. In much of today's world, it must be said, the same model prevails.

Enlightenment humanism, in contrast, explicitly rejects supernatural explanations as nothing more than the arbitrary, nescient, and error-laden lore of a pre-scientific age. It likewise dismisses divine authority as a relic of our species' intellectual infancy. Instead, humanists base their worldview on three broad Enlightenment pillars. Here, in a nutshell, is a sketch of each pillar.

REASON

The first and foremost pillar of Enlightenment humanism asserts the supremacy of reason and science as the foundations for knowledge. This is no idle claim, for reason champions a method - one absent from the lion's share of human history and possessed by no other philosophy - for reaching thoughtful, credible, and defendable conclusions. A method that demands skepticism, critical thought, and rational inquiry as the first steps on the path to knowledge. One that requires sound hypotheses that permit alternative outcomes. One that stipulates the formulation of detailed, testable predictions. One that calls for empirical evidence and the dispassionate evaluation of results. And one that in no instance relies on the fame, charisma, or authority of the claimant.

From this method, reason commands an unrivaled ability to demarcate likely truths from wishful convictions. Such truths, once verified, become universal. Said differently, a Muslim can verify a discovery by a Sikh, an atheist can corroborate a finding by a Jew, and a Taoist can confirm a claim by a Christian. Knowledge gained from reason thus imparts a universal veracity wholly unadulterated by religious affiliation, political inclination, or cultural sentiment. No other philosophy can assert a similar claim.

ETHICS

The second pillar affirms that we, *Homo sapiens*, are the product of evolution, a process that brought forth a social species with an innate moral compass. From reason and experience, secular societies fine-tune the compass to shape societal norms. The norms, like the societies they regulate, evolve over time.

The outcome of this process is no societal anomie, no dystopia of moral chaos. At worst, secular societies develop moral obligations and moral prohibitions that are indistinguishable from dogmatic belief systems. At best, such societies inspirit a social justice well beyond the arbitrary dictates of faith-based constructs. Indeed, countries whose inhabitants express low levels of religious belief - Denmark and Sweden, for example - routinely score highest on international measures of social equality, societal stability, and happiness.[1]

DEMOCRACY

Enlightenment humanism champions democracy as the governmental model best suited to endow equal rights upon all members of society. And, once those rights are constituted, to protect them. Democratic states accomplish this goal by embedding within their legislative, judicial, institutional, educational, and ethical frameworks a range of ideals rarely found in autocratic regimes. The more notable principles include individual liberty, the rule of law, equal rights before the law, separation of church from state, an independent press, and the recognition that sovereignty resides with the people.

To many humanists, the most critical of all such rights is freedom of expression, which embraces freedom to inquire, freedom to critique, and freedom to ridicule. For it is from this basic license that other freedoms emerge. While citizens in free societies worldwide enjoy this birthright, a multitude of theocracies and secular dictatorships continue to block - even condemn - similar liberties.

Reason. Ethics. Democracy. All three pillars underlie the promise of Enlightenment humanism. A promise that extols the transparency and grandeur of a fact-based world; that marvels at the ineffable beauty of the cosmos and our ability to comprehend it; that stands in awe of the nobler virtues of our species; that grants license to dream and to realize those dreams; and that magnifies joy in family, friendships, and life itself. All free from the oppressive weight of theistic hegemony and the self-serving dictates of despotic regimes.

The assorted axioms, aphorisms, and insights that follow capture this promise. They likewise render acknowledgement, a benefit that derives less from the selected excerpts than from the authors themselves. It is they, the denizens of this text, who affirm that throughout history a wide assortment of notable individuals have shared a similar philosophical outlook. The Enlightenment humanist, as such, is not alone. He or she enjoys the company of many of the most intelligent and thoughtful people ever to have walked the planet.

MISCELLANY

This and subsequent chapters close with assorted miscellany. Here is the first nugget. The text contains 1,150 entries that span forty-plus centuries. The earliest quotation, from a Sumerian cuneiform tablet, dates from 2600 BCE. The newest, from an ebook, was composed in 2020. A breakdown by century appears in the following table.

Number of Quotations by Year.

Year	Quotes	Percent
Before 1 CE	38	3%
1 - 1499	35	3%
1500 - 1599	31	3%
1600 - 1699	65	6%
1700 - 1799	136	12%
1800 - 1899	388	34%
1900 - 1999	445	39%
2000 - Today	12	1%
	1,150	100%

No fewer than 694 authors voice the entries. Some contributed a single quotation, others more. Collectively, the authors reflect a rich assortment of nationalities, ethnicities, races, genders, belief systems, and political leanings. Not all subscribed to Enlightenment humanism, but all expressed thoughts aligned to at least one of its tenets. Note, too, that in a handful of instances, I peppered the cento with quotations from detractors. Here, the contrast in their thinking serves to illuminate and reinforce Enlightenment principles.

A final comment concerns the structure of this volume. The entries are loosely organized along the three pillars of Enlightenment humanism. Within each pillar, the quotations appear alphabetically by author. Every effort was made to trace each quotation to a trusted primary or secondary source.

Diagram 1.1 A synopsis of Enlightenment humanism.

Enlightenment Humanism Simplified

Enlightenment humanism is a philosophy of life that accords to our species the faculty, guided by no supernatural authority, to lead happy, ethical, and meaningful lives. The philosophy rests on three pillars.

Pillar	Principle	Benchmark
Reason	Objective truths exist. Reason provides the most reliable means to discover these truths and to distinguish verifiable facts from wishful thinking.	Claims that assert truth must articulate a reasonable set of empirical, testable, and falsifiable conditions, each to be confirmed or disproved from experiment.
Ethics	Morality, a product of evolution, is innate. The citizens of each society fine-tune this built-in, moral compass from their customs, laws, and aspirations.	Societal norms are not set in stone but evolve over time. From these norms, all individuals derive identical rights. No one occupies a position of special moral privilege.
Democracy	Power flows upwards from citizens to elected leaders. All members of society share equal rights, liberties, and duties under constituted law.	The legitimacy of a government rests solely on the ability of the governed to displace their leaders without violence, evinced by a history of regularly having done so.

ON REASON

The invention of the scientific method and science is, I'm sure we'll all agree, the most powerful intellectual idea, the most powerful framework for thinking and investigating and understanding and challenging the world around us that there is, and it rests on the premise that any idea is there to be attacked. If it withstands the attack then it lives to fight another day, and if it doesn't withstand the attack then down it goes. Religion

doesn't seem to work like that.[2]

DOUGLAS ADAMS (1952-2001) ENGLISH WRITER

Facts are stubborn things; and whatever may be our wishes, our inclinations, or the dictates of our passions, they cannot alter the states of facts and evidence.[3]

JOHN ADAMS (1735-1826) AMERICAN PRESIDENT

In those parts of the world where learning and science have prevailed, miracles have ceased; but in such parts of it as are barbarous and ignorant, miracles are still in vogue.[4]

ETHAN ALLEN (1738-1789) AMERICAN REVOLUTIONARY

During all the years of my fervent belief, I simply had no idea how many sorry and embarrassing passages there are in the scriptural record. I had encountered a few instances of critical literature, but it had largely bounced off the armored shell of my bias and loyalty to the Catholic Church.[5]

STEVE ALLEN (1921-2000) AMERICAN WRITER

If you pray for rain long enough, it does eventually rain. If you pray for floodwaters to abate, they eventually do. The same happens in the absence of prayer.[6]

STEVE ALLEN (1921-2000) AMERICAN WRITER

To you I'm an atheist; to God I'm the loyal opposition.[7]

WOODY ALLEN (1935-) AMERICAN FILMMAKER

It is the arrogance of the contemporary to assume that your knowledge is enough do to something.[8] [Observing that what we consider true today often differs tomorrow.]

BRYAN APPLEYARD (1951-) BRITISH JOURNALIST

The earth revolves about the sun.[9] [Articulating the theory of heliocentrism some eighteen centuries prior to Copernicus.]
ARISTARCHUS OF SAMOS (C.310-C.230 BCE) GREEK ASTRONOMER

We must suppose that in the soul there is something contrary to the rational principle, resisting and opposing it.[10]
ARISTOTLE (384-322 BCE) GREEK PHILOSOPHER

The freethinking of one age is the common sense of the next.[11]
MATTHEW ARNOLD (1822-1888) ENGLISH POET

It is almost impossible to exaggerate the proneness of the human mind to take miracles as evidence, and to seek for miracles as evidence.[12]
MATTHEW ARNOLD (1822-1888) ENGLISH POET

The personages of the Christian heaven and their conversations are no more matter of fact than the personages of the Greek Olympus and their conversations.[13]
MATTHEW ARNOLD (1822-1888) ENGLISH POET

To surrender to ignorance and call it God has always been premature, and it remains premature today.[14]
ISAAC ASIMOV (1920-1992) AMERICAN WRITER

I believe in evidence. I believe in observation, measurement, and reasoning, confirmed by independent observers. I'll believe anything, no matter how wild or ridiculous, if there is evidence for it. The wilder and more ridiculous something is, however, the firmer and more solid the evidence will have to be.[15]
ISAAC ASIMOV (1920-1992) AMERICAN WRITER

Science doesn't purvey absolute truth. Science is a mechan-

ism. It's a way of trying to improve your knowledge of nature. It's a system for testing your thoughts against the universe and seeing if they match. And this works, not just for the ordinary aspects of science, but for all of life.[16]

ISAAC ASIMOV (1920–1992) AMERICAN WRITER

[Belief in gods is a] worn out but once useful crutch in mankind's journey towards truth. We consider the time has come for that crutch to be abandoned. To say that "God made the world" is simply a more or less sophisticated way of saying that we don't understand how the universe originated. A god, in so far as it is anything, is an admission of ignorance.[17]

PETER ATKINS (1940–) ENGLISH CHEMIST

Religion is a private pursuit of truth, driven by internal revelation and guided by authority. Science is the public pursuit of truth, guided by publicly executed experiments under controlled conditions. Religion is largely a culturally conditioned journey of sentimental wishful thinking. Science is transnational and transcultural. Religion is based on authority. Science is a ceaseless search for flaws in authority. You must choose between the feather bed of religion, which appeases the fears of the weak and encourages false hopes, and the rack of science, which exposes the true heart of the world.[18]

PETER ATKINS (1940–) ENGLISH CHEMIST

Ignorance of numbers prevents us from understanding things that are set down in scripture.[19] [Here, Augustine ascribes divine meaning to 153, the number of fish caught by a Biblical fisherman.[20] To prove his assertion, he adds to the number of Commandments (10) the number of gifts from the Holy Spirit (7),[21] giving 17. He then sums all the integers from 1 through 17, yielding 153.[22] Thus is religious prophesy affirmed.]

AUGUSTINE OF HIPPO (354–430) ROMAN AFRICAN THEOLOGIAN

The existence of trousers proves that God meant us to be bipeds.[23] [Reflecting on the rigor of religious claims.]

DAVID AUSTIN (1935–2005) BRITISH CARTOONIST

A hypothesis or theory which cannot be, at least in principle, falsified by empirical observations and experiments does not belong to the realm of science.[24]

FRANCISCO J. AYALA (1934–) SPANISH-AMERICAN BIOLOGIST

The theist, like the moralist, may believe that his experiences are cognitive experiences, but, unless he can formulate his "knowledge" in propositions that are empirically verifiable, we may be sure that he is deceiving himself.[25]

A. J. AYER (1910–1989) BRITISH PHILOSOPHER

The corruption of philosophy by the mixing it up with superstition and theology, in whole or in parts, is most injurious.[26]

FRANCIS BACON (1561–1626) ENGLISH NATURAL PHILOSOPHER

The reverence for antiquity, and the authority of men who have been esteemed great in philosophy, have retarded men from advancing in science, and almost enchanted them.[27] [Commenting, in part, on unquestioned devotion to ancient myths.]

FRANCIS BACON (1561–1626) ENGLISH NATURAL PHILOSOPHER

Nor should we neglect to observe that natural philosophy has, in every age, met with a troublesome and difficult opponent: I mean superstition, and a blind and immoderate zeal for religion.[28]

FRANCIS BACON (1561–1626) ENGLISH NATURAL PHILOSOPHER

Read not to contradict and confute; nor to believe and take for granted; nor to find talk and discourse; but to weigh and consider.[29]

FRANCIS BACON (1561–1626) ENGLISH NATURAL PHILOSOPHER

Every scientific theory is a formulation for explaining why things in the world are the way we perceive them to be.[30]

PHILIP BALL (1962–) BRITISH WRITER

Scientists do not join hands every Sunday and sing "Yes gravity is real! I know gravity is real! I will have faith! I believe in my heart that what goes up, up, up must come down, down, down. Amen!" If they did, we would think they were pretty insecure about the concept.[31]

DANIEL BARKER (1949–) AMERICAN MUSICIAN

The Christian Church at an early period of its existence took over from the Jews beliefs as to the creation and early history of the world and as to the origin of man. Such beliefs, as every educated person knows full well, can no longer be accepted.[32] [A stark and none-too-common admission from a Protestant bishop.]

ERNEST WILLIAM BARNES (1874–1953) ENGLISH THEOLOGIAN

If a theory is a good one, a fairly explicit statement of the falsifying conditions can be made.[33]

C. DANIEL BATSON (1943–) AMERICAN PSYCHOLOGIST

Representation of the world, like the world itself, is the work of men; they describe it from their own point of view, which they confuse with absolute truth.[34]

SIMONE DE BEAUVOIR (1908–1986) FRENCH WRITER

You're expected to evolve your thinking. Nothing is static, nothing is final, everything is held provisionally.[35] [Commenting on science.]

JOCELYN BELL BURNELL (1943-) BRITISH ASTROPHYSICIST

Few men think, yet all have opinions.[36]

GEORGE BERKELEY (1685-1753) IRISH PHILOSOPHER

Prove all things.[37]

THE BIBLE

Nothing found.[38]

SEARCH RESULT FOR THE WORD "SCIENCE" IN THE BIBLE

Nothing found.[39]

SEARCH RESULT FOR THE WORD "LOGIC" IN THE BIBLE

Nothing found.[40]

SEARCH RESULT FOR THE WORD "RATIONAL" IN THE BIBLE

To dogmatism, the spirit of inquiry is the same as the spirit of evil.[41]

AMBROSE BIERCE (1842-C.1914) AMERICAN SATIRIST

Education: That which discloses to the wise and disguises from the foolish their lack of understanding.[42]

AMBROSE BIERCE (1842-C.1914) AMERICAN SATIRIST

Love reason then;
and let whatever you write,
borrow from her alone,
from her beauty and her light.[43]

NICOLAS BOILEAU (1636-1711) FRENCH POET

The greatest obstacle to discovery is not ignorance - it is the illusion of knowledge.[44]

DANIEL J. BOORSTIN (1914-2004) AMERICAN HISTORIAN

Primitive people are necessarily ignorant people; and ignorance breeds fear, and fear breeds gods.[45]

HYPATIA BRADLAUGH BONNER (1858-1935) BRITISH ACTIVIST

Ignorance, the foe of science, is the ally of religion.

SERGE BOURGAINE

In every instance where the Church has tried to offer its own description of the world and there was some scientific alternative on the very same topic, the latter has proved better. Every battle has been lost and conclusively so.[46]

PASCAL BOYER (?) FRENCH-AMERICAN ANTHROPOLOGIST

How can people believe such nonsense?[47] [Recalling the words of a Catholic theologian who derided a primitive tribe's belief in witches.]

PASCAL BOYER (?) FRENCH-AMERICAN ANTHROPOLOGIST

Ignorance is fatal.[48]

RAY BRADBURY (1920-2012) AMERICAN WRITER

Science demands that facts not be subordinated to opinions, but that opinions be subordinated to facts.[49]

BERTOLT BRECHT (1898-1956) GERMAN WRITER

Man masters nature not by force but by understanding. That is why science has succeeded where magic failed: because it has looked for no spell to cast on nature.[50]

JACOB BRONOWSKI (1908-1974) BRITISH MATHEMATICIAN

To explain the unknown by the known is a logical procedure; to explain the known by the unknown is a form of theological lunacy.[51]

DAVID MARSHALL BROOKS (?) AMERICAN HUMANIST

I have heard an atheist defined as a man who had no invisible means of support.[52]

JOHN BUCHAN (1875–1940) SCOTTISH WRITER

Those who would legislate against the teaching of evolution should also legislate against gravity, electricity and the ... velocity of light.[53]

LUTHER BURBANK (1849–1926) AMERICAN BOTANIST

If we take science as our sole guide, if we accept and hold fast that alone which is verifiable, the old theology, with all its miraculous machinery, must go.[54]

JOHN BURROUGHS (1837–1921) AMERICAN NATURALIST

Some people speak as if we were not justified in rejecting a theological doctrine unless we can prove it false. But the burden of proof does not lie upon the rejecter.[55]

JOHN BAGNELL BURY (1861–1927) IRISH HISTORIAN

That we shall probably never know the whole "truth" about the universe does not really matter very much; the fun comes in trying to find out.[56]

NIGEL CALDER (1931–2014) BRITISH WRITER

Christian philosophy bids [reason] give place, and yield complete submission to the Holy Spirit.[57] [One of endless attempts to subordinate reason to faith.]

JOHN CALVIN (1509–1564) FRENCH THEOLOGIAN

What gods are there, what gods have there ever been, that were not from man's imagination?[58]

JOSEPH CAMPBELL (1904-1987) AMERICAN SCHOLAR

Multiple hypotheses should be proposed whenever possible. Proposing alternative explanations that can answer a question is good science. If we operate with a single hypothesis, especially one we favor, we may direct our investigation toward a hunt for evidence in support of this hypothesis.[59]

NEIL CAMPBELL (1946-2004) AMERICAN BIOLOGIST

I was pretty good in science. But again, because of the small budget, in science class we couldn't do experiments in order to prove theories. We just believed everything. Actually I think that class was called Religion. Religion was always an easy class. All you had to do was suspend the logic and reasoning you were taught in all the other classes.[60]

GEORGE CARLIN (1937-2008) AMERICAN COMEDIAN

It is a mathematical fact that the casting of a pebble from my hand alters the centre of gravity of the universe.[61]

THOMAS CARLYLE (1795-1881) SCOTTISH PHILOSOPHER

The aim of science is to discover and illuminate truth.[62]

RACHEL CARSON (1907-1964) AMERICAN SCIENTIST

A scientist and a theologian are out for a drive in the country.
Theologian: Oh, look! Those sheep have been shorn.
Scientist: Yes, on this side.[63]
[On assumption versus exactitude in the formation of knowledge.]

THOMAS CATHCART (1940-) AND DANIEL KLEIN (1939-) AMERICAN PHILOSOPHERS

There are two kinds of truth; the truth that lights the way and the truth that warms the heart. The first of these is science, and the second is art.[64]

RAYMOND CHANDLER (1888-1959) AMERICAN-BRITISH WRITER

It is not commonly understood that intellectual liberty rests on doubt.[65]

ÉMILE CHARTIER (1868-1951) FRENCH PHILOSOPHER

One experiment is not enough to confirm a hypothesis, but a single contradictory result suffices to reject it.[66]

ÉMILIE DU CHÂTELET (1706-1749) FRENCH SCIENTIST

There are some people - and I am one of them - who think that the most practical and important thing about a man is his view of the universe.[67]

G. K. CHESTERTON (1874-1936) ENGLISH WRITER

There is a noticeable general difference between the sciences and mathematics on the one hand, and the humanities and social sciences on the other. It's a first approximation, but one that is real. In the former, the factors of integrity tend to dominate more over the factors of ideology. It's not that scientists are more honest people. It's just that nature is a harsh taskmaster. You can lie or distort the story of the French Revolution as long as you like, and nothing will happen. Propose a false theory in chemistry, and it'll be refuted tomorrow.[68]

NOAM CHOMSKY (1928-) AMERICAN PHILOSOPHER

For disgrace lies not in imperfect knowledge, but in foolish and obstinate continuance in a state of imperfect knowledge.[69]

CICERO (106-43 BCE) ROMAN STATESMAN

The fact is that the most useful parts of science have been investigated for the sake of truth, and not for their usefulness.[70]

WILLIAM KINGDON CLIFFORD (1845–1879) ENGLISH MATHEMATICIAN

Scientific thought does not mean thought about scientific subjects with long names. There are no scientific subjects. The subject of science is the human universe; that is to say, everything that is, or has been, or may be related to man.[71]

WILLIAM KINGDON CLIFFORD (1845–1879) ENGLISH MATHEMATICIAN

Gods are fragile things, they may be killed by a whiff of science or a dose of common sense.[72]

CHAPMAN COHEN (1868–1954) ENGLISH SECULARIST

Science is not satisfied with psychological certitude, for the mere intensity with which a belief is held is no guarantee of its truth.[73]

MORRIS COHEN (1880–1947) AND ERNEST NAGEL (1901–1985) AMERICAN PHILOSOPHERS

The first man of science was he who looked into a thing, not to learn whether it furnished him with food, or shelter, or weapons, or tools, or ornaments, or playwiths, but who sought to know it for the gratification of knowing.[74]

SAMUEL TAYLOR COLERIDGE (1772–1834) ENGLISH PHILOSOPHER

A man ceases to be a beginner in any given science and becomes a master in that science when he has learned that he is going to be a beginner all his life.[75]

ROBIN COLLINGWOOD (1889–1943) ENGLISH PHILOSOPHER

We owe almost all our knowledge, not to those have agreed, but to those who have differed.[76]

CHARLES COLTON (1780-1832) ENGLISH CLERIC

It is not so difficult a task to plant new truths, as to root out old errors.[77]

CHARLES COLTON (1780-1832) ENGLISH CLERIC

The most Illustrious Lord Cardinal Bellarmine, in the name of His Holiness the Pope and the whole Congregation of the Holy Office, ordered and enjoined Galileo to abandon completely the opinion that the sun stands still and the earth moves, and henceforth not to hold, teach or defend it in any way whatever.[78] [The first conviction of Galileo.]

CONGREGATION OF CARDINALS (26 FEBRUARY 1616) CATHOLIC CHURCH

For holding as true the false doctrine that the sun is the center of the world and the earth moves; for being in correspondence with some German mathematicians about it; and for interpreting Holy Scripture according to your own meaning - we condemn you to formal imprisonment.[79] [The second conviction of Galileo.]

CONGREGATION OF CARDINALS (22 JUNE 1633) CATHOLIC CHURCH

Free thought, free speech and free criticism are the life of science.[80]

EDWIN GRANT CONKLIN (1863-1952) AMERICAN BIOLOGIST

In the middle of all stands the sun.[81] [Arguably the most famous scientific challenge to theistic authority.]

NICOLAUS COPERNICUS (1473-1543) PRUSSIAN ASTRONOMER

If perchance there should be foolish speakers who, together with those ignorant of all mathematics, will take it upon themselves to [find] some place in the scriptures wickedly distorted

to their purpose, [and] should dare to assail this my work, they are of no importance to me, to such an extent do I despise their judgment as rash.[82] [Copernicus censuring his critics in a letter to Pope Paul III.]

NICOLAUS COPERNICUS (1473–1543) PRUSSIAN ASTRONOMER

It's entirely wrong, and it's the road back to the cave. The way we got out of the caves and into modern civilization is through the process of understanding and thinking. Those things were not done by gut instinct.[83] [Commenting on public mistrust of professional expertise.]

BRIAN COX (1968–) ENGLISH PHYSICIST

In religion, faith is a virtue. In science, faith is a vice.[84]

JERRY COYNE (1949–) AMERICAN BIOLOGIST

I am among those who think that science has great beauty.[85]

MARIE CURIE (1867–1934) POLISH PHYSICIST

To guide the human spirit to its noble destination, to its knowledge of truth. To spread healthy ideas to [all] classes of people. To remove men from the influence of prejudice and passion. To establish reason as the arbiter and the supreme guide of public opinion. This is the essential object of science. This is how science contributes to the advancement of civilization.[86]

GEORGES CUVIER (1769–1832) FRENCH NATURALIST

It has often and confidently been asserted, that man's origin can never be known, but ignorance more frequently begets confidence than does knowledge: it is those who know little, and not those who know much, who so positively assert that this or that problem will never be solved by science.[87]

CHARLES DARWIN (1809–1882) ENGLISH NATURALIST

It has become almost a cliché to remark that nobody boasts of ignorance of literature, but it is socially acceptable to boast of ignorance of science and proudly proclaim incompetence in mathematics.[88]

RICHARD DAWKINS (1941-) ENGLISH EVOLUTIONARY BIOLOGIST

Not a single fossil contradicts evolution.[89]

RICHARD DAWKINS (1941-) ENGLISH EVOLUTIONARY BIOLOGIST

Even if not a single fossil had ever been found, the evidence from surviving animals would still overwhelmingly force the conclusion that Darwin was right.[90]

RICHARD DAWKINS (1941-) ENGLISH EVOLUTIONARY BIOLOGIST

For what each man wishes, that he also believes to be true.[91]

DEMOSTHENES (384-322 BCE) GREEK STATESMAN

Darwinism is laying down to die. Ample printer's ink yet will be wasted before its demise is complete, and before it no longer serves as a weapon against Christianity.[92] [Hoping, in vain, for a single, glorious discovery that will upend the theory of natural selection, solely to preserve a favored, mystical fiction.]

EBERHARD DENNERT (1861-1942) GERMAN NATURALIST

In due course, the Darwinian Revolution will come to occupy a ... secure and untroubled place in the minds - and hearts - of every educated person on the globe.[93]

DANIEL DENNETT (1942-) AMERICAN COGNITIVE SCIENTIST

Biology is engineering.[94]

DANIEL DENNETT (1942-) AMERICAN COGNITIVE SCIENTIST

Design revision in nature must follow the profligate method of releasing and test-driving many variants and letting the losers die, unexamined.[95] [Affirming a harsh but confirmed consequence of natural selection.]

DANIEL DENNETT (1942-) AMERICAN COGNITIVE SCIENTIST

In order to examine into the truth, it is necessary once in one's life to doubt of all things.[96]

RENÉ DESCARTES (1596-1650) FRENCH PHILOSOPHER

It is not enough to have a good mind. The main thing is to use it well.[97]

RENÉ DESCARTES (1596-1650) FRENCH PHILOSOPHER

I think, therefore I am not a Christian.[98] [A variant on Descartes' maxim: *I think, therefore I am.*[99]

KARLHEINZ DESCHNER (1924-2014) GERMAN WRITER

When the standard for acceptance of scientific theories was their compatibility with the Bible, science not only stagnated but relinquished what it had gained. The standing of science declined to the same degree as that of the Bible advanced.[100]

KARLHEINZ DESCHNER (1924-2014) GERMAN WRITER

Every great advance in science has issued from a new audacity of imagination.[101]

JOHN DEWEY (1859-1952) AMERICAN PHILOSOPHER

To maintain the state of doubt and to carry on systematic and protracted inquiry - these are the essentials of thinking.[102]

JOHN DEWEY (1859-1952) AMERICAN PHILOSOPHER

Reality is that which, when you stop believing in it, doesn't go away.[103]

PHILIP K. DICK (1928–1982) AMERICAN WRITER

Take nothing on its looks; take everything on evidence. There's no better rule.[104]

CHARLES DICKENS (1812–1870) ENGLISH WRITER

Skepticism is the first step towards truth.[105]

DENIS DIDEROT (1713–1784) FRENCH PHILOSOPHER

The history of science is not a mere record of isolated discoveries; it is a narrative of the conflict of two contending powers, the expansive force of the human intellect on one side, and the compression arising from traditionary faith and human interests on the other.[106]

JOHN WILLIAM DRAPER (1811–1882) AMERICAN SCIENTIST

[Science is] a series of judgments, revised without ceasing.[107]

ÉMILE DUCLAUX (1840–1904) FRENCH MICROBIOLOGIST

Agreement with experiment is the sole criterion of truth for a physical theory.[108]

PIERRE DUHEM (1861–1916) FRENCH PHYSICIST

Truth is a stern companion; she bristles with iron; she wounds those whom she attacks, and sometimes those who speak to her.[109]

ALEXANDRE DUMAS (1802–1870) FRENCH WRITER

Science is not a sacred cow. Science is a horse. Don't worship it. Feed it.[110]

ABBA EBAN (1915–2002) ISRAELI DIPLOMAT

I think it is not irreligion but a tidiness of mind which rebels against the idea of permeating scientific research with a religious implication.[111]

ARTHUR EDDINGTON (1882-1944) ENGLISH PHYSICIST

[What] is too absurd for fiction is only to be found in the sober pages of science.[112]

ARTHUR EDDINGTON (1882-1944) ENGLISH PHYSICIST

The truth of a theory can never be proven, for one never knows if future research will contradict its conclusions.[113]

ALBERT EINSTEIN (1879-1955) GERMAN PHYSICIST

Science can flourish only in an atmosphere of free speech.[114]

ALBERT EINSTEIN (1879-1955) GERMAN PHYSICIST

When science progresses, it often opens vaster mysteries to our gaze. Moreover, science frequently discovers that it must abandon or modify what it once believed.[115]

LOREN EISELEY (1907-1977) AMERICAN ANTHROPOLOGIST

Charles Robert Darwin ... taught us that we can understand life's history in purely naturalistic terms, without recourse to the supernatural or divine.[116]

NILES ELDRIDGE (1943-) AMERICAN ANTHROPOLOGIST

Standard procedure in scientific experiments calls for planning data collection and statistical tests in advance of actually collecting the data, in order to avoid "cherry-picking" - the deprecated practice of examining a data set after the fact and then figuring out what statistical test will produce a significant result of the desired form.[117]

WESLEY ELSBERRY (1960-) AMERICAN DATA SCIENTIST

We would be 1,500 years ahead if it hadn't been for the church dragging science back by its coattails and burning our best minds at the stake.[118]

CATHERINE FAHRINGER (1922-2008) AMERICAN ACTIVIST

You'll be riding along in an automobile, you'll be the driver, perhaps; you're a Christian. There will be several other people in the automobile with you, maybe someone who is not a Christian. When the trumpet sounds, you and other born-again believers in the automobile will be instantly caught away. You will disappear, leaving behind only your clothing and physical things that cannot inherit eternal life. That unsaved person ... in the automobile will be startled to find that the car is moving along without a driver, and suddenly somewhere crashes. These saved people in the car have disappeared.[119] [One of countless religious yarns concocted to sustain a cherished myth, here describing the rapture.]

JERRY FALWELL (1933-2007) AMERICAN TELEVANGELIST

The first principle is that you must not fool yourself - and you are the easiest person to fool.[120]

RICHARD FEYNMAN (1918-1988) AMERICAN PHYSICIST

It doesn't seem to me that this fantastically marvelous universe, this tremendous range of time and space and different kinds of animals, and all the different planets, and all these atoms with all their motions, and so on, all this complicated thing can merely be a stage so that God can watch human beings struggle for good and evil - which is the view that religion has. The stage is too big for the drama.[121]

RICHARD FEYNMAN (1918-1988) AMERICAN PHYSICIST

It does not make any difference how smart you are, who made the guess, or what his name is - if it disagrees with experiment it is wrong.[122]

RICHARD FEYNMAN (1918–1988) AMERICAN PHYSICIST

Science transcends all times and ages.[123]

JOHANN GOTTLIEB FICHTE (1762–1814) GERMAN PHILOSOPHER

No man has ever sat down calmly unbiased to reason out his religion, and not ended by rejecting it.[124]

HENRY FIELDING (1707–1754) ENGLISH WRITER

Inductive inference is the only process known to us by which essentially new knowledge comes into the world.[125]

RONALD FISHER (1890–1962) BRITISH STATISTICIAN

We need go back only a few centuries to find the great mass of people depending on religion for the satisfaction of practically all their wishes. From rain out of the sky to good health on earth, they sought their desires at the altars of their gods. Whether they wanted large families, good crops, freedom from pestilence, or peace of mind, they conceived themselves as dependent on the favor of heaven. Then science came with its alternative, competitive method of getting what we want. That is science's most important attribute. As an intellectual influence it is powerful enough, but as a practical way of achieving man's desires it is overwhelming.[126]

HARRY EMERSON FOSDICK (1878–1969) AMERICAN PASTOR

Unreason is to reason as dazzlement is to daylight.[127]

MICHEL FOUCAULT (1926–1984) FRENCH PHILOSOPHER

Either the pope must abolish knowledge and printing, or printing at length will root him out.[128] [Commenting on the recent invention of the printing press and its ability to disseminate knowledge, thus "rooting out" religious superstitions.]

JOHN FOXE (C.1516-1587) ENGLISH HISTORIAN

The wonder is not that the field of the stars is so vast, but that man has measured it.[129]

ANATOLE FRANCE (1844-1924) FRENCH WRITER

The way to see by faith is to shut the eye of reason.[130]

BENJAMIN FRANKLIN (1706-1790) AMERICAN STATESMAN

Science and everyday life cannot and should not be separated.[131]

ROSALIND FRANKLIN (1920-1958) AMERICAN SCIENTIST

I see no reason to believe that a creator of protoplasm or primeval matter, if such there be, has any reason to be interested in our insignificant race in a tiny corner of the universe, and still less in us, as still more insignificant individuals.[132]

ROSALIND FRANKLIN (1920-1958) AMERICAN SCIENTIST

[Religious] theories are those which you and many other people find easiest and pleasantest to believe, but, so far as I can see, they have no foundation other than they lead to a pleasant view of life and an exaggerated idea of our own importance.[133] [Franklin, to her father.]

ROSALIND FRANKLIN (1920-1958) AMERICAN SCIENTIST

No diminishment of science can alter the fact that it seeks to comprehend the real world, while religion is an illusion, one that derives its strength by satiating our instinctual desires.[134]

SIGMUND FREUD (1856-1939) AUSTRIAN NEUROLOGIST

They [religious teachings] are all illusions, unprovable.[135]

SIGMUND FREUD (1856-1939) AUSTRIAN NEUROLOGIST

You will hear things like, "Science doesn't know everything." Well, of course science doesn't know everything. But because science doesn't know everything, that doesn't mean science knows nothing.[136]

STEPHEN FRY (1957-) BRITISH WRITER

The man of science dissects the statement, tests the facts, and demonstrates order, even where he cannot [fully disclose] its purpose.[137]

MARGARET FULLER (1810-1850) AMERICAN JOURNALIST

In questions of science, the authority of a thousand is not worth the humble reasoning of a single individual.[138]

GALILEO GALILEI (1564-1642) ITALIAN PHYSICIST

Philosophy is written in this grand book, the universe, which stands continually open to our gaze. But the book cannot be understood unless one first learns to comprehend the language and read the letters in which it is composed. It is written in the language of mathematics.[139]

GALILEO GALILEI (1564-1642) ITALIAN PHYSICIST

I should think that anyone who considered it more reasonable for the whole universe to move in order to let the Earth remain fixed would be more irrational than one who should climb to the top of a cupola just to get a view of the city and its environs, and then demand that the whole countryside should revolve around him so that he would not have to take the trouble to turn his head.[140]

GALILEO GALILEI (1564-1642) ITALIAN PHYSICIST

In the long run, good science drives out bad.[141]

MARTIN GARDNER (1914-2010) AMERICAN WRITER

Science is humble. It knows what it knows, and it knows what it doesn't know. It bases its conclusions and beliefs on hard evidence - evidence that is constantly updated and upgraded. It doesn't get offended when new facts come along. It doesn't hold on to medieval practices because they are tradition.[142]

RICKY GERVAIS (1961-) ENGLISH COMEDIAN

Is it not the believer who is rigid in his faith? Is not the theist staunch in his unshakable faith in the existence of an Almighty? We rationalists, on the contrary, are ready to change, ready to modify our views whenever we find reason. The only thing we can possibly be rigid about is our conviction based on reason, on truth.[143]

PRABIR GHOSH (1945-) INDIAN WRITER

While [Darwinian] man, however well behaved,
At best is only a monkey shaved.[144]

W. S. GILBERT (1836-1911) AND ARTHUR SULLIVAN (1842-1900) ENGLISH DRAMATISTS

We are so constituted that we believe the most incredible things: and once they are firmly engraved upon the memory, woe to him who would attempt to eradicate them.[145]

JOHANN WOLFGANG VON GOETHE (1749-1832) GERMAN WRITER

In times of general ignorance, the prevailing tone of thought is ... religious.[146]

CHARLES GORHAM (1812-1901) AMERICAN DIPLOMAT

Male mammals have nipples because females need them.[147]

[Explaining how evolution creates non-functional body parts; in this instance, by noting that it is simpler to direct the epicene growth of an embryo from a single set of DNA code.]

STEPHEN J. GOULD (1941-2002) AMERICAN PALEONTOLOGIST

The organism proposes and the environment disposes.[148] [Abridging the theory of natural selection.]

STEPHEN J. GOULD (1941-2002) AMERICAN PALEONTOLOGIST

The fundamentalists, by "knowing" the answers before they start, and then forcing nature into the straitjacket of their discredited preconceptions, lie outside the domain of science - or of any honest intellectual inquiry.[149]

STEPHEN J. GOULD (1941-2002) AMERICAN PALEONTOLOGIST

Cows now ruminate where women once prayed - a notable advance.[150] [Remarking on a convent converted into a cattle farm.]

REMY DE GOURMONT (1858-1915) FRENCH POET

Misuse of reason might yet return the world to pre-technological night; plenty of religious zealots hunger for just such a result, and are happy to use the latest technology to effect it.[151]

A. C. GRAYLING (1949-) BRITISH PHILOSOPHER

I am putting together a secular bible. My Genesis is when the apple falls on Newton's head.[152]

A. C. GRAYLING (1949-) BRITISH PHILOSOPHER

The human race's favorite method for being in control of facts is to ignore them.[153]

CELIA GREEN (1935-) BRITISH WRITER

It is axiomatic that the importance of a given thing cannot be determined by the number of people who are prepared to admit that it is important.[154]

CELIA GREEN (1935–) BRITISH WRITER

So perfect and complete was my ignorance that it was impossible for me to know that I knew nothing.[155] [A gap to be filled, one hopes, from education, not indoctrination.]

HANS JAKOB CHRISTOFFEL VON GRIMMELSHAUSEN (C.1621–1676) GERMAN WRITER

Seeking knowledge is a duty upon every Muslim.[156]

THE HADITH

My practice as a scientist is atheistic. That is to say, when I set up an experiment I assume that no god, angel, or devil is going to interfere with its course; and this assumption has been justified by such success as I have achieved in my professional career. I should therefore be intellectually dishonest if I were not also atheistic in the affairs of the world.[157]

JOHN HALDANE (1892–1964) BRITISH SCIENTIST

From the fact that there are 400,000 species of beetles on this planet, but only 8,000 species of mammals, [one might] conclude that the Creator, if He exists, has a special preference for beetles.[158]

JOHN HALDANE (1892–1964) BRITISH SCIENTIST

317 is a prime, not because we think so, or because our minds are shaped in one way rather than another, but because it is so, because mathematical reality is built that way.[159]

G. H. HARDY (1877–1947) BRITISH MATHEMATICIAN

We have to repudiate this idea that beliefs can be sanctified by something other than evidence and argument.[160]

SAM HARRIS (1967-) AMERICAN NEUROSCIENTIST

Words like "God" and "Allah" must go the way of "Apollo" and "Baal".[161]

SAM HARRIS (1967-) AMERICAN NEUROSCIENTIST

Non-rational opinions and beliefs appear to their possessor as propositions whose truth is immediately obvious, and whose validity is silly to question or to doubt.[162]

BERNARD HART (1879-1966) BRITISH PSYCHIATRIST

I have noticed that even people who claim that everything is predestined, and that we can do nothing to change it, look before they cross the road.[163]

STEPHEN HAWKING (1942-2018) ENGLISH PHYSICIST

We are just a slightly advanced breed of monkeys on a minor planet that circles a very average star. But we can understand the universe, and that makes us something very special.[164]

STEPHEN HAWKING (1942-2018) ENGLISH PHYSICIST

A lot of prizes have been awarded for showing the universe is not as simple as we might have thought.[165]

STEPHEN HAWKING (1942-2018) ENGLISH PHYSICIST

My goal is simple. It is a complete understanding of the universe, why it is as it is and why it exists at all.[166]

STEPHEN HAWKING (1942-2018) ENGLISH PHYSICIST

Whosoever studies works of science must, if he wants to find the truth, transform himself into a critic of everything he reads.

He must examine tests and explanations with the greatest precision and question them from all angles and aspects.[167]

HASAN IBN AL-HAYTHAM (C.965-C.1040) ARAB SCIENTIST

The origin of all science is in the desire to know causes; and the origin of all false science and imposture is in the desire to accept false causes rather than none; or, which is the same thing, in the unwillingness to acknowledge our own ignorance.[168]

WILLIAM HAZLITT (1778-1830) ENGLISH WRITER

There is a natural tendency in sects to narrow the mind.[169]

WILLIAM HAZLITT (1778-1830) ENGLISH WRITER

Science embraces facts and debates opinions; religion embraces opinions and debates facts.[170]

TOM HEEHLER (?) AMERICAN WRITER

Where one burns books, one eventually burns people.[171] [A prescient observation from one whose books were burned by the Nazis.]

HEINRICH HEINE (1797-1856) GERMAN WRITER

One can judge from experiment, or one can blindly accept authority.[172]

ROBERT HEINLEIN (1907-1988) AMERICAN WRITER

Beware of the truth, gentle Sister. Although much sought after, truth can be dangerous to the seeker. Myths and reassuring lies are much easier to find and believe. If you find a truth, even a temporary one, it can demand that you make painful changes.[173]

FRANK HERBERT (1920-1986) AMERICAN WRITER

A mind exclusively bent upon the idea of utility necessarily narrows the range of the imagination.[174]

JOHN HIBBEN (1861-1933) AMERICAN PHILOSOPHER

[Science is] an imaginative adventure of the mind seeking truth in a world of mystery.[175]

CYRIL HERMAN HINSHELWOOD (1897-1967) ENGLISH CHEMIST

A common fallacy in much of the adverse criticism to which science is subjected today is that it claims certainty, infallibility and complete emotional objectivity. It would be more nearly true to say that it is based upon wonder, adventure and hope.[176]

CYRIL NORMAN HINSHELWOOD (1897-1967) ENGLISH CHEMIST

What can be asserted without evidence can also be dismissed without evidence.[177]

CHRISTOPHER HITCHENS (1949-2011) ENGLISH-AMERICAN WRITER

Gullibility and credulity are considered undesirable qualities in every department of human life - except religion.[178]

CHRISTOPHER HITCHENS (1949-2011) ENGLISH-AMERICAN WRITER

The first cause of absurd conclusions I ascribe to the want of Method.[179]

THOMAS HOBBES (1588-1679) ENGLISH PHILOSOPHER

The central task of education is to implant a will and a facility for learning; it should produce not learned but learning people.[180]

ERIC HOFFER (1922-1983) AMERICAN PHILOSOPHER

A good catchword can obscure analysis for fifty years.[181] [Or, in the case of Christianity, 2000 years.]

OLIVER WENDELL HOLMES, JR. (1841-1935) AMERICAN JURIST

There is nothing more deceptive than an obvious fact.[182]

SHERLOCK HOLMES (C.1854-C.1930) FICTIONAL DETECTIVE

I have all the coin types mentioned in the Bible, confirming that such denominations were real. The historical part of the Bible is therefore trustworthy.[183] [Erecting a fallacy of composition to assert the truth of a whole from the truth of a part.]

EDMOND HOLROYD (?) ATMOSPHERIC SCIENTIST

It still remains true that as a set of cognitive beliefs, religion is a speculative hypothesis of an extremely low order of probability.[184]

SIDNEY HOOK (1902-1989) AMERICAN PHILOSOPHER

The admission that the existence of God cannot be demonstrated is often coupled with the retort that neither can his non-existence be demonstrated - as if this puts belief and disbelief on an equally reasonable footing. If we are unjustified in disbelieving an assertion save only when its contradictory is demonstrated to be impossible, we should have to believe that the universe is populated with the wildest fantasies.[185]

SIDNEY HOOK (1902-1989) AMERICAN PHILOSOPHER

Sapere aude - Dare to be wise.[186] [An epigram later revitalized by Kant.]

HORACE (65-68 BCE) ROMAN POET

In science it is evidence that counts, not opinion.[187]

SABINE HOSSENFELDER (1976-) GERMAN PHYSICIST

Rational arguments don't usually work on religious people.

Otherwise there would be no religious people.[188]
HOUSE (2007)

Science is understood, while theology is believed.[189]
ELBERT HUBBARD (1856-1915) AMERICAN WRITER

Equipped with his five senses, man explores the universe around him and calls the adventure science.[190]
EDWIN HUBBLE (1889-1953) AMERICAN ASTRONOMER

In each village in France there is now a lighted torch, the schoolmaster, and a mouth which blows upon it, the priest.[191] [Commenting on the nationwide introduction of public education in nineteenth-century France.]
VICTOR HUGO (1802-1885) FRENCH NOVELIST

The dogmatic views of past centuries live on only in the prejudices of people who, conscious of their weakness, like to cover themselves in darkness.[192]
ALEXANDER VON HUMBOLDT (1769-1859) PRUSSIAN NATURALIST

Errors in religion are dangerous; those in philosophy only ridiculous.[193]
DAVID HUME (1711-1776) BRITISH PHILOSOPHER

No testimony is sufficient to establish a miracle, unless that testimony be of such kind that its falsehood would be more miraculous than the fact which it endeavours to establish.[194]
DAVID HUME (1711-1776) BRITISH PHILOSOPHER

If we reason "a priori," anything may appear able to produce anything.[195]
DAVID HUME (1711-1776) BRITISH PHILOSOPHER

Facts do not cease to exist because they are ignored.[196]

ALDOUS HUXLEY (1894–1963) BRITISH WRITER

If the idea of an external personality is already consciously or unconsciously at the back of our minds, we shall almost inevitably mold our experience in accordance with that idea.[197]

JULIAN HUXLEY (1887–1975) ENGLISH BIOLOGIST

The great tragedy of science is the slaying of a beautiful hypothesis by an ugly fact.[198]

THOMAS HUXLEY (1825–1895) ENGLISH BIOLOGIST

Scepticism is the highest of duties; blind faith the one unpardonable sin.[199]

THOMAS HUXLEY (1825–1895) ENGLISH BIOLOGIST

The man of science has learned to believe in justification, not by faith but by verification.[200]

THOMAS HUXLEY (1825–1895) ENGLISH BIOLOGIST

Infidels are intellectual discoverers. They sail the unknown seas and find new isles and continents in the infinite realms of thought. An infidel is one who has found a new fact, who has an idea of his own, and who in the mental sky has seen another star. He is an intellectual capitalist, and for that reason excites the envy and hatred of the theological pauper.[201]

ROBERT GREEN INGERSOLL (1833–1899) AMERICAN LAWYER

As people become more intelligent they care less for preachers and more for teachers.[202]

ROBERT GREEN INGERSOLL (1833–1899) AMERICAN LAWYER

Every fact is an enemy of the church.[203]

ROBERT GREEN INGERSOLL (1833–1899) AMERICAN LAWYER

This mindless tolerance, which places observable scientific facts, subject to proof, on the same level as unprovable supernatural fantasy, has played a major role in the resurgence of both anti-intellectualism and anti-rationalism.[204]

SUSAN JACOBY (1945–) AMERICAN AUTHOR

Whatever universe a professor believes in must at any rate be a universe that lends itself to lengthy discourse. A universe definable in two sentences is something for which the professorial intellect has no use.[205] [Reflecting on declarations of faith as explanatory paradigms.]

WILLIAM JAMES (1842–1910) AMERICAN PHILOSOPHER

First ... a new theory is attacked as absurd; then it is admitted to be true, but obvious and insignificant; finally, it is seen to be so important that its adversaries claim that they themselves have discovered it.[206]

WILLIAM JAMES (1842–1910) AMERICAN PHILOSOPHER

The priests of the different religious sects dread the advance of science as witches do the approach of daylight.[207]

THOMAS JEFFERSON (1743–1826) AMERICAN PRESIDENT

We are not afraid to follow truth wherever it may lead, nor to tolerate any error so long as reason is left free to combat it.[208]

THOMAS JEFFERSON (1743–1826) AMERICAN PRESIDENT

Ridicule is the only weapon which can be used against unintelligible propositions. Ideas must be distinct before reason can act upon them.[209] [Here, from context, Jefferson equates "unintelligible propositions" to theology.]

THOMAS JEFFERSON (1743-1826) AMERICAN PRESIDENT

Science literacy isn't about figuring out how to solve equations like $e=mc^2$. Rather, it's about being able to read an article in the newspaper about the environment, about health care, and figuring out how to vote on it. It's about being able to prepare nutritious meals. It's about being able to think your way through the day.[210]

MAE JEMISON (1956-) AMERICAN ENGINEER

I would trade all of my technology for an afternoon with Socrates.[211]

STEVE JOBS (1955-2011) AMERICAN ENTREPRENEUR

No one can claim to be immune to the spirit of his own epoch or to possess anything like a complete knowledge of it.[212]

CARL JUNG (1875-1961) SWISS PSYCHIATRIST

A reliable way to make people believe in falsehoods is frequent repetition, because familiarity is not easily distinguished from truth.[213]

DANIEL KAHNEMAN (1934-) ISRAELI-AMERICAN PSYCHOLOGIST

Our comforting conviction that the world makes sense rests on a secure foundation: our almost unlimited ability to ignore our ignorance.[214]

DANIEL KAHNEMAN (1934-) ISRAELI-AMERICAN PSYCHOLOGIST

There are many examples of old, incorrect theories that stubbornly persisted, sustained only by the prestige of foolish but well-connected scientists. ... Many of these theories have been killed off only when some decisive experiment exposed their incorrectness.[215] [On the power of experiment to quash deeply

held but erroneous convictions, a phenomenon well-recognized in science but decried in religion.]
MICHIO KAKU (1947-) AMERICAN PHYSICIST

Sapere aude - Dare to think.[216] [Expressing the maxim of the Enlightenment.]
IMMANUEL KANT (1724-1804) GERMAN PHILOSOPHER

Theology depends on a double standard. One set of standards is employed for reading and interpreting one's own tradition and its texts; another, for the texts and traditions of all other.[217]
WALTER KAUFMANN (1921-1980) GERMAN-AMERICAN PHILOSOPHER

For the great enemy of truth is very often not the lie - deliberate, contrived and dishonest - but the myth - persistent, persuasive, and unrealistic. We subject all facts to a prefabricated set of interpretations.[218]
JOHN KENNEDY (1917-1963) AMERICAN PRESIDENT

Without proper experiments I conclude nothing.[219]
JOHANNES KEPLER (1571-1630) GERMAN SCIENTIST

In theology we must consider the weight of authority; in science, the weight of reason.[220]
JOHANNES KEPLER (1571-1630) GERMAN SCIENTIST

The modern scientific method is non-negotiable in its all-encompassing importance as a worldview. Many would argue it is the only worldview that a rational, thinking person can have in explaining how and why the world is the way it is.[221]
JIM AL-KHALILI (1962-) BRITISH PHYSICIST

Everything must be doubted.[222]

SØREN KIERKEGAARD (1813-1855) DANISH PHILOSOPHER

Reason stops where religion begins.[223]

MIKE KIMPS (1950-) AMERICAN SAGE

The beauty of religious mania is that it has the power to explain everything. Once God (or Satan) is accepted as the first cause of everything which happens in the moral world, nothing is left to chance ... or change.[224]

STEPHEN KING (1947-) AMERICAN WRITER

It is ironic that several of these individuals, who so staunchly and proudly touted their religious convictions in public, would time and again lie to cover their tracks and disguise the real purpose behind the Intelligent Design Policy.[225] [Remarks from the presiding judge on school board members who attempted to have creationism taught in public schools.]

KITZMILLER V DOVER AREA SCHOOL DISTRICT (2005) UNITED STATES DISTRICT COURT

One of the most persistent fallacies about the Christian Church is that it kept learning alive during the Dark and Middle Ages. What the Church did was to keep learning alive in the monasteries, while preventing the spread of knowledge outside them.[226]

MARGARET K. KNIGHT (1903-1983) ENGLISH PSYCHOLOGIST

Men are superstitious in proportion as they are ignorant.[227]

RICHARD PAYNE KNIGHT (1751-1824) ENGLISH CLASSICAL SCHOLAR

Every atom in your body was once inside a star that exploded. Moreover, the atoms in your left hand probably came from a different star than did those in your right hand. We are all, liter-

ally, star children, and our bodies made of stardust.[228]
LAWRENCE KRAUSS (1954-) AMERICAN-CANADIAN PHYSICIST

I believe that the pursuit of truth and right ideas through honest debate and rigorous argument is a noble undertaking.[229]
CHARLES KRAUTHAMMER (1950-2018) AMERICAN JOURNALIST

Skepticism is an antidote for nonsense.[230]
PAUL KURTZ (1925-2012) AMERICAN PHILOSOPHER

There is nothing more innately human than the tendency to transmute what has become customary into what has been divinely ordained.[231]
SUZANNE LA FOLLETTE (1893-1983) AMERICAN WRITER

God, once imagined to be an omnipresent force throughout the whole world of nature and man, has been increasingly tending to seem omni-absent. Everywhere intelligent and educated people rely more and more on purely secular and scientific techniques for the solution of their problems. As science advances, belief in divine miracle and the efficacy of prayer becomes fainter and fainter.[232]
CORLISS LAMONT (1902-1995) AMERICAN PHILOSOPHER

Never believe anything that can't be proved.[233]
IRVING LANGMUIR (1881-1957) AMERICAN CHEMIST

Knowledge is often a sore impediment to the movements of common sense.[234]
PETER MERE LATHAM (1789-1875) ENGLISH PHYSICIAN

We ought, in every instance, to submit our reasoning to the test of experiment, and never to search for truth but by the nat-

ural road of experiment and observation.[235]

ANTOINE LAVOISIER (1743-1794) FRENCH CHEMIST

[Miracles] multiply or diminish in proportion to the intensity with which the imagination is directed to theological topics.[236]

W. E. H. LECKY (1838-1903) IRISH HISTORIAN

Physics is not a religion. If it were, we'd have a much easier time raising money.[237]

LEON LEDERMAN (1922-) AMERICAN PHYSICIST

Get in line in that processional,
Step into that small confessional,
There, the guy who's got religion'll
Tell you if your sin's original.
If it is, try playin' it safer,
Drink the wine and chew the wafer,
Two, four, six, eight,
Time to transubstantiate![238]

TOM LEHRER (1928-) AMERICAN SATIRIST

The scientific mind does not so much provide the right answers as ask the right questions.[239]

CLAUDE LÉVI-STRAUSS (1908-2009) FRENCH ANTHROPOLOGIST

There exists a type of transcendental ventriloquism that can make men believe that something said on earth comes from Heaven.[240]

GEORG CHRISTOPH LICHTENBERG (1742-1799) GERMAN SCIENTIST

When all think alike, no one thinks very much.[241]

WALTER LIPPMANN (1889-1974) AMERICAN WRITER

To love truth, for truth's sake, is the principal part of human perfection in this world, and the seed-plot of all other virtues.[242]

JOHN LOCKE (1632–1704) ENGLISH PHILOSOPHER

False or doubtful positions, relied upon as unquestioned maxims, keep those in the dark from truth.[243]

JOHN LOCKE (1632–1704) ENGLISH PHILOSOPHER

Every sect, as far as reason will help them, gladly use it; when it fails them, they cry out it is a matter of faith, and above reason.[244]

JOHN LOCKE (1632–1704) ENGLISH PHILOSOPHER

Sophistry is only fit to make men more conceited in their ignorance.[245]

JOHN LOCKE (1632–1704) ENGLISH PHILOSOPHER

Reason must be our last judge and guide in everything.[246]

JOHN LOCKE (1632–1704) ENGLISH PHILOSOPHER

Whoever wants to be a Christian must be intent on silencing the voice of reason.[247] [Observing, correctly, that reason is a threat to dogmatic beliefs.]

MARTIN LUTHER (1483–1546) GERMAN THEOLOGIAN

Christ wants to slay reason.[248]

MARTIN LUTHER (1483–1546) GERMAN THEOLOGIAN

Reason is the greatest whore of the devil.[249] [An ironic attack on reason from someone who sought to escape the choke hold of the Catholic Church.]

MARTIN LUTHER (1483–1546) GERMAN THEOLOGIAN

A Christian of the fifth century with a Bible is neither better nor worse situated than a Christian of the nineteenth century with a Bible.[250] [On those who profess knowledge and certitude from a chimerical text.]

THOMAS BABINGTON MACAULAY (1800–1859) BRITISH HISTORIAN

Vulgar minds are more impressed by example than by reason.[251]

MACROBIUS AMBROSIUS THEODOSIUS (C.1193–1280) GREEK WRITER

REASON

For it is [the task] of natural science not simply to accept what we are told but to inquire into the causes of natural things.[252]

ALBERTUS MAGNUS (C.1193–1280) GERMAN THEOLOGIAN

Religion ... stops people from thinking because they think all the answers are in that one book.[253]

BILL MAHER (1956) AMERICAN COMEDIAN

Reason conquers all.[254]

MARCUS MANILIUS (CIRCA 100 CE) ROMAN POET

Fanaticism turns into a means of salvation, enthusiasm into epileptic ecstasy, politics becomes an opiate for the masses, a proletarian eschatology; and reason veils her face.[255] [Commenting on ideological extremism.]

THOMAS MANN (1875–1955) GERMAN WRITER

I count religion but a childish toy, and hold there is no sin but ignorance.[256]

CHRISTOPHER MARLOWE (1564–1593) ENGLISH WRITER

The world holds two classes of men; intelligent men without religion, and religious men without intelligence.[257]

ABUL 'ALA AL-MA'ARRI (973–1057) ARAB PHILOSOPHER

Reason is to the philosopher what grace is to the Christian.[258]

CÉSAR CHESNEAU DU MARSAIS (1676–1756) FRENCH GRAMMARIAN

They must find it hard to take truth for authority who have so long mistaken authority for truth.[259]

GERALD MASSEY (1828–1907) ENGLISH EGYPTOLOGIST

The modern clergyman has acquired in his study of science ... an astonishing facility for explaining things away.[260]

W. SOMERSET MAUGHAM (1874–1965) ENGLISH WRITER

Metaphysics is the finding of bad reasons for what we believe upon instinct.[261]

W. SOMERSET MAUGHAM (1874–1965) ENGLISH WRITER

In every branch of knowledge the progress is proportional to the amount of facts on which to build.[262]

JAMES CLERK MAXWELL (1831–1879) SCOTTISH PHYSICIST

The theist and the scientist are rival interpreters of nature. The one retreats as the other advances.[263]

JOSEPH MCCABE (1867–1955) ENGLISH WRITER

There is no idea or belief I so dearly cherish so as to shield it from rigorous scrutiny or thoughtful challenge. There is no idea or belief I esteem so highly that I will not alter it or abandon it - sacrifice it in favor of standing even closer to the truth.[264]

CARLESPIE MARY ALICE MCKINNEY (?) AMERICAN WRITER

The invisible and the non-existent look very much alike.[265]

DELOS MCKOWN (1930-) AMERICAN PHILOSOPHER

Children must be taught how to think, not what to think.[266]

MARGARET MEAD (1901-1978) AMERICAN ANTHROPOLOGIST

The purpose of scientific enquiry is not to compile an inventory of factual information, nor to build up a totalitarian world picture of natural laws in which every event that is not compulsory is forbidden. We should think of it rather as a logically articulated structure of justifiable beliefs about nature.[267]

PETER MEDAWAR (1915-1987) BRAZILIAN-BRITISH BIOLOGIST

The intensity of the conviction that a hypothesis is true has no bearing on whether it is true or not.[268]

PETER MEDAWAR (1915-1987) BRAZILIAN-BRITISH BIOLOGIST

Science makes people reach selflessly for truth and objectivity; it teaches people to accept reality, with wonder and admiration, not to mention the deep awe and joy that the natural order of things brings to the true scientist.[269]

LISE MEITNER (1878-1868) AUSTRIAN-SWEDISH PHYSICIST

Science, at bottom, is really anti-intellectual. It always distrusts pure reason, and demands the production of objective fact.[270]

H. L. MENCKEN (1880-1956) AMERICAN JOURNALIST

The principle itself of dogmatic religion, dogmatic morality, dogmatic philosophy, is what requires to be rooted out; not any particular manifestation of that principle.[271]

JOHN STUART MILL (1806-1873) BRITISH PHILOSOPHER

He who knows only his own side of the case, knows little of that.[272]

JOHN STUART MILL (1806-1873) BRITISH PHILOSOPHER

Miracles have no claim whatever to the character of historical facts and are wholly invalid as evidence of any revelation.[273]

JOHN STUART MILL (1806-1873) BRITISH PHILOSOPHER

The beauty of mathematics only shows itself to more patient followers.[274]

MARYAM MIRZAKHANI (1977-2017) IRANIAN-AMERICAN MATHEMATICIAN

The criterion of truth is that it works even if nobody is prepared to acknowledge it.[275]

LUDWIG VON MISES (1881-1973) AUSTRIAN ECONOMIST

Reason at its best shuns all extremes.[276]

MOLIÈRE (1622-1673) FRENCH PLAYWRIGHT

Science begins where bigotry and absolute certainty end.[277]

ASHLEY MONTAGU (1905-1999) BRITISH-AMERICAN ANTHROPOLOGIST

Whoever wishes to be cured of ignorance must first confess it.[278]

MICHEL DE MONTAIGNE (1533-1592) FRENCH PHILOSOPHER

Man is certainly crazy. He could not make a mite, [but] he makes gods by the dozen.[279]

MICHEL DE MONTAIGNE (1533-1592) FRENCH PHILOSOPHER

We give the name scientist to the type of man who has felt experiment to be a means guiding him to search out the deep truth of life, to lift a veil from its fascinating secrets, and who, in this pursuit, has felt arising within him a love for the mysteries of nature, so passionate as to annihilate the thought of himself.[280]

MARIA MONTESSORI (1870-1952) ITALIAN EDUCATOR

Crowd: Hail Messiah!
Brian: I'm not the Messiah. Will you please listen? I am not the Messiah, do you understand? Honestly!
Woman: Only the true Messiah denies His divinity.
Brian: What?[281] [Dumbfounded.]

MONTY PYTHON (1979) BRITISH COMEDIANS

Emperor Napoleon I: *Monsieur Laplace, they tell me you have written this large book [Système du Monde] on the system of the universe, and have never even mentioned its Creator.*
Pierre-Simon Laplace: *I did not have need of that hypothesis.*[282]

REPORTED BY AUGUSTUS DE MORGAN (1806-1871) BRITISH MATHEMATICIAN

The only way we can determine the true age of the earth is for God to tell us what it is.[283] [Expressing, beautifully, the antithesis of reason.]

HENRY MORRIS (1918-2006) AMERICAN CREATIONIST

It is ... the unexplored that attracts us.[284]

MURASAKI SHIKIBU (C.973-C.1031) JAPANESE WRITER

This sort of claptrap was good for the Stone Age, when people actually believed that if they prayed for rain they would get it. But we're a grown-up world now, and it's time to put away childish things.[285]

MADALYN MURRAY (1919-1995) AMERICAN ACTIVIST

Knowledge is the death of research.[286]

WALTHER NERNST (1864-1941) GERMAN CHEMIST

Plato is my friend, Aristotle is my friend, but truth is my greatest friend.[287]

ISAAC NEWTON (1642-1726) ENGLISH POLYMATH

Men insist most vehemently upon their certainties when their hold upon them has been shaken.[288]

RHEINHOLD NIEBUHR (1892-1971) AMERICAN THEOLOGIAN

We possess art lest we perish of the truth.[289]

FRIEDRICH NIETZSCHE (1844-C.1900) GERMAN PHILOSOPHER

Belief in truth begins by doubting all truths that one has previously believed.[290]

FRIEDRICH NIETZSCHE (1844-C.1900) GERMAN PHILOSOPHER

Faith means not wanting to know what is true.[291]

FRIEDRICH NIETZSCHE (1844-C.1900) GERMAN PHILOSOPHER

If you want to deny evolution and live in your world that's completely inconsistent with everything we observe in the universe, that's fine, but don't make your kids do it.[292]

BILL NYE (1955-) AMERICAN EDUCATOR

What would it take to change your mind?
Response from science: *Evidence.*
Response from creationism: *Nothing.*[293]

BILL NYE (1955-) AMERICAN EDUCATOR AND KEN HAM (1951-) AMERICAN CREATIONIST

Logic is the most useful tool of all the arts. Without it no science can be fully known.[294]

WILLIAM OF OCKHAM (C.1287–1347) ENGLISH THEOLOGIAN

In science, you don't just get to cherry-pick the parts that justify what you were going to do anyway. That's religion. You're thinking of religion.[295] [Reacting to a talk show guest who advised listeners to believe only the scientific studies that one wishes to believe.]

JOHN OLIVER (1977–) BRITISH SATIRIST

Every breath you draw, every accelerated beat of your heart in the emotional periods of your oratory depend upon highly elaborated physical and chemical reactions and mechanisms which nature has been building up through a million centuries. ... Not only this, but some of your highest ideals of human fellowship and comradeship were not created in a moment, but represent the work of ages.[296] [Osborn, responding to a creationist orator.]

HENRY FAIRFIELD OSBORN (1857–1935) AMERICAN PALEONTOLOGIST

Wilberforce [derisively]: *Is it through your grandfather or grandmother that you claim descent from a monkey?*
Huxley: *I am not ashamed to have a monkey for an ancestor; but I would be ashamed to be connected with a man who used great gifts to obscure the truth.*[297]
[An insightful but likely apocryphal interchange.]

OXFORD EVOLUTION DEBATE (1860) BETWEEN ENGLISH BISHOP SAMUEL WILBERFORCE (1805–1873) AND ENGLISH BIOLOGIST THOMAS HUXLEY (1825–1895)

Person A: *Creationism should be taught in school.*
Person B: *In a music class?*
Person A: *Creationism is not music.*

Person B: *Neither is it science.*
OVERHEARD BY THE EDITOR

To argue with a man who has renounced the use and authority of reason ... is like administering medicine to the dead.[298]
THOMAS PAINE (1737–1809) ENGLISH ACTIVIST

It is error only, and not truth, that shrinks from inquiry.[299]
THOMAS PAINE (1737–1809) ENGLISH ACTIVIST

One good schoolmaster is of more use than a hundred priests.[300]
THOMAS PAINE (1737–1809) ENGLISH ACTIVIST

Man, when it is useful to him, can profess faith in a theory of which he knows nothing more than its name; a phenomenon common in all religions.[301]
VILFREDO PARETO (1848–1903) ITALIAN ECONOMIST

Is the truth depressing? Some may find it so. But I find it liberating, and consoling.[302]
DEREK PARFIT (1942–2017) BRITISH PHILOSOPHER

Truth is one form of loveliness.[303]
THEODORE PARKER (1810–1860) AMERICAN CLERIC

Judge by argument the much disputed proof uttered by me.[304] [An early admonition not to blindly accept assertions from authority.]
PARMENIDES OF ELEA (C.500 BCE) GREEK PHILOSOPHER

Reason commands us far more imperiously than a master; when we disobey the latter we are unhappy, when we disobey

the former we are fools.[305]
BLAISE PASCAL (1623–1662) FRENCH MATHEMATICIAN

A scientist can be productive in various ways. One is having the ability to plan and carry out experiments, but the other is having the ability to formulate new ideas, which can be about what experiments can be carried out ... by making [the] proper calculations. Individual scientists who are successful in their work are successful for different reasons.[306]
LINUS PAULING (1901–1994) AMERICAN CHEMIST

We must not forget that some of the best ideas seemed like nonsense at first. The truth will prevail in the end. Nonsense will fall of its own weight.[307]
CECILIA PAYNE-GAPOSCHKIN (1900–1979) BRITISH-AMERICAN ASTRONOMER

It is the man of science, eager to have his every opinion regenerated, his every idea rationalized, by drinking at the fountain of fact, and devoting all the energies of his life to the cult of truth, not as he understands it, but as he does not yet understand it, that ought properly to be called a philosopher.[308]
CHARLES SANDERS PEIRCE (1839–1914) AMERICAN PHILOSOPHER

The first rule of reason: do not block the way of inquiry.[309]
CHARLES SANDERS PEIRCE (1839–1914) AMERICAN PHILOSOPHER

Inquiry is human;
blind obedience brutal.
Truth never loses by the one,
but often suffers by the other.[310]
WILLIAM PENN (1644–1718) ENGLISH COLONIST

Decision [often] is the fruit of ignorance; hesitation, of reflec-

tion.[311]

PERICLES (C.495-429 BCE) ATHENIAN STATESMAN

Perhaps the greatest discovery in human history is that our traditional sources of belief are in fact generators of error and should be dismissed as grounds for knowledge.[312]

STEVEN PINKER (1954-) AMERICAN COGNITIVE PSYCHOLOGIST

Understanding that our understanding might be wrong is essential, and trying to figure out the ways we may be mistaken is the only way that science can help us find our way to the truth.[313]

PHILIP PLAIT (1964-) AMERICAN ASTRONOMER

An experiment is a question which science poses to nature, and a measurement is the recording of nature's answer.[314]

MAX PLANCK (1858-1947) GERMAN PHYSICIST

The unexamined life is not worth living.[315]

PLATO (C.428-C.348 BCE) GREEK PHILOSOPHER

Knowledge is the food of the soul.[316]

PLATO (C.428-C.348 BCE) GREEK PHILOSOPHER

We prove from logic, but we discover from intuition.[317]

HENRI POINCARÉ (1854-1912) FRENCH MATHEMATICIAN

Experiment is the sole source of truth. It alone can teach us something new; it alone can give us certainty.[318]

HENRI POINCARÉ (1854-1912) FRENCH MATHEMATICIAN

Justice therefore forbids, and reason itself forbids, the State to be godless; or to ... treat the various religions ... alike, and to bestow upon them promiscuously equal rights and privileges.[319] [A sordid attempt, under the pretext of justice and reason, to maintain religious authority.]

POPE LEO XIII (1810-1903)

If unbridled license of speech and of writing be granted to all, nothing will remain sacred and inviolate.[320] [So?]

POPE LEO XIII (1810-1903)

Submission of mind and will must be shown ... in such a way that [the Roman Pontiff] is acknowledged with reverence, the judgments made by him are sincerely adhered to, according to his manifest mind and will.[321] [Note the similarity to proclamations by despots in North Korea, Iran, and elsewhere.]

POPE PAUL VI (1897-1978)

Science is one of the very few human activities - perhaps the only one - in which errors are systematically criticized and fairly often, in time, corrected.[322]

KARL POPPER (1902-1994) AUSTRIAN-BRITISH PHILOSOPHER

Only through the growth of our knowledge can we liberate our minds from their spiritual enslavement: enslavement by prejudices, idols, and avoidable errors.[323]

KARL POPPER (1902-1994) AUSTRIAN-BRITISH PHILOSOPHER

It must be possible for an empirical scientific system to be refuted by experience.[324] [Affirming the critical importance of testable and falsifiable hypotheses.]

KARL POPPER (1902-1994) AUSTRIAN-BRITISH PHILOSOPHER

It is not his possession of knowledge, of irrefutable truth, that makes the man of science, but his persistent and recklessly crit-

ical quest for truth.[325]

KARL POPPER (1902-1994) AUSTRIAN-BRITISH PHILOSOPHER

Whenever religion has had rule over law, the result has historically always been an automatic violation of human rights and an attack on factual education.[326]

ARON RA (1962-) AMERICAN ACTIVIST

Theism is nothing more than a bunch of fables made up by superstitious primitives. Having a doctorate in theology is consequently not significantly different than having a Ph.D. in Mother Goose.[327]

ARON RA (1962-) AMERICAN ACTIVIST

Science is a great game. It is inspiring and refreshing. The playing field is the universe itself.[328]

ISIDOR ISAAC RABI (1898-1988) AMERICAN PHYSICIST

Wisdom lies in thinking. The spear-head of thinking is rationalism.[329]

E. V. RAMASAMY AKA PERIYAR (1879-1973) INDIAN ACTIVIST

Religion ... is the first enemy of the ability to think.[330]

AYN RAND (1905-1982) AMERICAN WRITER

Atheists face facts without fear.[331]

GOPARAJU RAMACHANDRA RAO, AKA GORA (1902-1975) INDIAN ACTIVIST

What passes as eternal truth is usually uncritical assumption and hasty generalization erected into unquestionable first principles.[332] [On faith.]

SIDNEY RATNER (1908-1996) AMERICAN ECONOMIST

Shall reason or prejudice be the arbiter of our destiny?[333] [Here, from context, "prejudice" is a synonym for religion.]
SIDNEY RATNER (1908-1996) AMERICAN ECONOMIST

The essence of religion is inertia; the essence of science is change. It is the function of the one to preserve, it is the function of the other to improve.[334]
WINWOOD READE (1838-1875) BRITISH HISTORIAN

The curses and spells used in the books are actual curses and spells; which when read by a human being risk conjuring evil spirits into the presence of the person reading the text.[335] [Explaining a ban on Harry Potter novels at a Catholic school.]
DAN REEHIL (?-) CATHOLIC PRIEST

None of the miracles with which old stories are filled took place under scientific conditions. Miracles happen only in times and countries where they already are believed, and before persons predisposed to believe them.[336]
ERNEST RENAN (1823-1892) FRENCH PHILOLOGIST

For indignation is so easy and satisfying a mood that it is apt to prevent one from attending to any facts that oppose it.[337]
LEWIS FRY RICHARDSON (1881-1953) ENGLISH MATHEMATICIAN

Questions are what drive science, not answers.[338]
ROBERT ROOT-BERNSTEIN (1953-) AMERICAN BIOLOGIST

I encourage [my students] to be skeptical - as long as their skepticism is based on logic and evidence.[339]
ROBERT ROOT-BERNSTEIN (1953-) AMERICAN BIOLOGIST

Take nothing for granted, I counsel my students: that is what makes a scientist.[340]

ROBERT ROOT-BERNSTEIN (1953-) AMERICAN BIOLOGIST

If children understood how to reason, they would not need to be educated.[341]

JEAN-JACQUES ROUSSEAU (1712-1778) FRENCH PHILOSOPHER

The Royal Society's motto ... is an expression of the determination of Fellows to withstand the domination of authority and to verify all statements by an appeal to facts determined by experiment.[342]

THE ROYAL SOCIETY

Error is not a mere accident of an untrained intellect, but a necessary stage [in] the expression of the truth.[343]

JOSIAH ROYCE (1855-1916) AMERICAN PHILOSOPHER

Science progresses best when observations force us to alter our preconceptions.[344]

VERA RUBIN (1928-2016) AMERICAN ASTRONOMER

[My faith], which was never strong, is being beaten into mere gold leaf. If only the geologists would let me alone, I could do very well, but those dreadful hammers! I hear the clink of them at the end of every Bible verse.[345] [Musing on the displacement of religious dogma by scientific truths; in this instance, that geologists found planet Earth to be far older than its purported Biblical age.]

JOHN RUSKIN (1819-1900) ENGLISH POLYMATH

To modern educated people, it seems obvious that matters of fact are to be ascertained by observation, not by consulting ancient authorities.[346]

BERTRAND RUSSELL (1872-1970) BRITISH PHILOSOPHER

Throughout the last 400 years, during which the growth of science had gradually shown men how to acquire knowledge of the ways of nature and mastery over natural forces, the clergy have fought a losing battle against science, in astronomy and geology, in anatomy and physiology, in biology and psychology and sociology. Ousted from one position, they have taken up another. After being worsted in astronomy, they did their best to prevent the rise of geology; they fought against Darwin in biology, and at the present time they fight against scientific theories of psychology and education. At each stage, they try to make the public forget their earlier obscurantism, in order that their present obscurantism may not be recognized for what it is.[347]

BERTRAND RUSSELL (1872-1970) BRITISH PHILOSOPHER

No opinion should be with fervour. No one holds with fervour that seven times eight is fifty-six, because it can be shown to be the case. Fervour is only necessary in commending an opinion which is doubtful or demonstrably false.[348]

BERTRAND RUSSELL (1872-1970) BRITISH PHILOSOPHER

Nowadays, only the wilfully ignorant dismiss the truth that we evolved from earlier ancestors.[349]

ADAM RUTHERFORD (1975-) BRITISH GENETICIST

Scientists are not dependent on the ideas of a single man, but on the combined wisdom of thousands of men, all thinking of the same problem, and each doing his little bit to add to the great structure of knowledge which is gradually being erected.[350]

ERNEST RUTHERFORD (1871-1937) NEW ZEALAND PHYSICIST

All science is either physics or stamp collecting.[351]
ERNEST RUTHERFORD (1871-1937) NEW ZEALAND PHYSICIST

Science is much more than a body of knowledge. It is a way of thinking.[352]
CARL SAGAN (1934-1996) AMERICAN PHYSICIST

[Religious] hypotheses are often framed precisely so they are invulnerable to any experiment that offers a prospect of disproof, so even in principle they cannot be invalidated.[353]
CARL SAGAN (1934-1996) AMERICAN PHYSICIST

There isn't a religion on the planet that doesn't long for a comparable ability - precise, and repeatedly demonstrated before committed skeptics - to foretell future events.[354] [A splendid accolade to the scientific method.]
CARL SAGAN (1934-1996) AMERICAN PHYSICIST

If you want to save your child from polio, you can pray or you can inoculate.[355]
CARL SAGAN (1934-1996) AMERICAN PHYSICIST

Extraordinary claims require extraordinary evidence.[356]
CARL SAGAN (1934-1996) AMERICAN PHYSICIST

It works. It is not perfect. It can be misused. It is only a tool. But it is by far the best tool we have, self-correcting, ongoing, applicable to everything. It has two rules. First: there are no sacred truths; all assumptions must be critically examined; arguments from authority are worthless. Second: whatever is inconsistent with the facts must be discarded or revised. We must understand the cosmos as it is and not confuse how it is with how we wish it to be.[357] [On the nature of science.]

CARL SAGAN (1934–1996) AMERICAN PHYSICIST

If you wish to make an apple pie from scratch, you must first invent the universe.[358]

CARL SAGAN (1934–1996) AMERICAN PHYSICIST

Skepticism is the chastity of the intellect, and it is shameful to surrender it too soon or to the first comer.[359]

GEORGE SANTAYANA (1863–1952) SPANISH PHILOSOPHER

My atheism, like that of Spinoza, is true piety towards the universe and denies only gods fashioned by men in their own image, to be servants of their human interests.[360]

GEORGE SANTAYANA (1863–1952) SPANISH PHILOSOPHER

It is difficult to discriminate the voice of truth from the clamour raised by heated partisans.[361]

FRIEDRICH SCHILLER (1759–1805) GERMAN PHILOSOPHER

Martyrdom has always been proof of the intensity, never the correctness of a belief.[362]

ARTHUR SCHNITZLER (1862–1931) AUSTRIAN WRITER

A certain amount of general ignorance is the condition of all religions, the element in which alone they can exist.[363]

ARTHUR SCHOPENHAUER (1788–1860) GERMAN PHILOSOPHER

Faith and knowledge behave like two scales of a balance: as one goes up, the other goes down.[364]

ARTHUR SCHOPENHAUER (1788–1860) GERMAN PHILOSOPHER

Religion always has and always will clash with the unbiased, noble pursuit of pure truth.[365]

ARTHUR SCHOPENHAUER (1788-1860) GERMAN PHILOSOPHER

Reason wishes the decision that it gives to be just; anger wishes to have the decision which it has given seem the just decision.[366]

SENECA THE YOUNGER (C.4 BCE-65 CE) ROMAN PHILOSOPHER

Dogma: Assent to the non-evident.[367]

SEXTUS EMPIRICUS (C.160-C.210 CE) GREEK OR ROMAN PHYSICIAN

Every why hath a wherefore.[368]

WILLIAM SHAKESPEARE (1564-1616) ENGLISH PLAYWRIGHT

Learning learns but one lesson: doubt![369]

GEORGE BERNARD SHAW (1856-1950) IRISH PLAYWRIGHT

Nothing stands between the people and [religious] fictions except the silly falsehood that the fictions are literal truths.[370]

GEORGE BERNARD SHAW (1856-1950) IRISH PLAYWRIGHT

The fact that a believer [might be] happier than a skeptic is no more to the point than the fact that a drunken man [might be] happier than a sober one.[371]

GEORGE BERNARD SHAW (1856-1950) IRISH PLAYWRIGHT

That which is incapable of proof itself is no proof of anything else.[372]

PERCY BYSSHE SHELLEY (1792-1822) ENGLISH POET

Science begins with the null hypothesis: Your claim isn't true until proven otherwise.[373]

MICHAEL SHERMER (1954-) AMERICAN WRITER

I have heard many times that atheists know more about religion than religious people. Atheism is an effect of that knowledge, not a lack of knowledge. I gave a Bible to my daughter. That's how you make atheists.[374]

DAVID SILVERMAN (1966-) AMERICAN ACTIVIST

It is often said that knowledge is power, but it might be more correct to say that "thinking" is power.[375] [From context, the author refers to critical thinking.]

STUART SIM (?-) ENGLISH WRITER

I do not think evolution is supremely important because it is my specialty. On the contrary, it is my specialty because I think it is supremely important.[376]

GEORGE GAYLORD SIMPSON (1902-1984) AMERICAN PALEONTOLOGIST

The knowledge of anything, since all things have causes, is not acquired or complete unless it is known by its causes.[377]

IBN SINA (C.980-1037) PERSIAN SCIENTIST

Beginning to reason is like stepping onto an escalator that leads upward and out of sight. Once we take the first step, the distance to be traveled is independent of our will and we cannot know in advance where we shall end.[378]

PETER SINGER (1946-) AUSTRALIAN PHILOSOPHER

Science is the great antidote to the poison of enthusiasm and superstition.[379]

ADAM SMITH (1723-1790) SCOTTISH PHILOSOPHER

Those who regard the fundamentalist revival as a harmless return to religious values would do well to take a closer look. There is a deep, underlying revolt here - a rejection of the ra-

tionalism and humanism of modern society.[380]
GEORGE SMITH (1949-) AMERICAN WRITER

It is the nature of science that once a position becomes orthodox, it should be subjected to criticism.[381] [A principle foreign to authoritarian ideologies, both political and religious.]
JOHN MAYNARD SMITH (1920-2004) BRITISH GENETICIST

Ideas, as the raw material from which knowledge is produced, exist in superabundance, but that makes the production of knowledge more difficult rather than easier.[382]
THOMAS SOWELL (1930-) AMERICAN ECONOMIST

Those who cavalierly reject the Theory of Evolution, as not adequately supported by facts, seem quite to forget that their own theory is supported by no facts at all.[383]
HERBERT SPENCER (1820-1903) ENGLISH PHILOSOPHER

[Religion] requires nothing but obedience.[384]
BENEDICT DE SPINOZA (1632-1677) DUTCH PHILOSOPHER

Prophets have a peculiar faculty for imagination but no special gift of understanding.[385] [Observing that revealed wisdom merely conforms to the preconceived notions of the advocate.]
BENEDICT DE SPINOZA (1632-1677) DUTCH PHILOSOPHER

The scope of philosophy [is] nothing but truth; that of faith nothing but piety and obedience.[386]
BENEDICT DE SPINOZA (1632-1677) DUTCH PHILOSOPHER

I object to intellect without discipline.[387]
VOICED BY THE CHARACTER SPOCK (1967) STAR TREK

An error becomes an error when born as truth.[388]

LEC STANISLAW (1909–1966) POLISH POET

Create your own myths. That's how the gods started.[389]

LEC STANISLAW (1909–1966) POLISH POET

The terror of the thunderstorm led primitive man to the conception of a Supreme Being whose attribute was the thunderbolt. But when Franklin brought the lightning from the clouds and showed it to be a mere electric spark, when we learned to make the lightning harmless by the lightning-rod, and when finally we harnessed electricity to do our work, naturally our reverence for the thrower of the thunderbolt decayed. So the gods of experience vanished.[390]

CHARLES PROTEUS STEINMETZ (1865–1923) GERMAN-AMERICAN ENGINEER

Religion robs one of the courage to contemplate unusual things and condemns self-examination as the most egregious of all sins.[391]

STENDHAL (1783–1842) FRENCH WRITER

The universe is not fine-tuned to us; we are fine-tuned to our particular universe.[392]

VICTOR J. STENGER (1935–2014) AMERICAN PHYSICIST

Science is not going to change its commitment to the truth. We can only hope religion changes its commitment to nonsense.[393]

VICTOR J. STENGER (1935–2014) AMERICAN PHYSICIST

What we see is beautiful; what we know is more beautiful; what we don't know is by far the most beautiful.[394]

NICOLAS STENO (1638–1686) DANISH SCIENTIST

The scientific method is deliberately constructed to overcome a deep-seated human wish to assume that what we want to be true - what we claim to "know"- really is true.[395]

IAN STEWART (1945-) BRITISH MATHEMATICIAN

Given their virtually seamless record of error when it comes to pontificating about scientific discoveries, you'd think Roman Catholic bishops might have acquired a little humility by now.[396]

THOMAS SUTCLIFFE (?-) BRITISH JOURNALIST

I would rather have questions I can't answer than answers I can't question.[397]

MAX TEGMARK (1967-) SWEDISH-AMERICAN COSMOLOGIST

Court: *What do you find?*

Foreman: *We have found for the state, found the defendant guilty.*

Court: *Mr. Scopes, the jury has found you guilty under this indictment, in violation of what is commonly known as the anti-evolution statute, which makes it unlawful for any teacher to teach in any of the public schools of the state, any theory that denies the story of the divine creation of man, and teach instead thereof that man has descended from a lower order of animals.*[398]

[The verdict in a 1925 trial where a high school teacher was found guilty of violating a law that prohibited the teaching of evolution in state-funded schools.]

TENNESSEE V JOHN THOMAS SCOPES (1925)

Truth and hatred of truth start together.[399]

TERTULLIAN (C.155-C.240 CE) CARTHAGINIAN WRITER

A multitude of words is no proof of a prudent mind.[400]

THALES OF MILETUS (C.624-C.545 BCE) GREEK PHILOSOPHER

[Water is] that of which all things consist, the first from which they come to be, the last into which they are resolved, this is the element and this the principle of things, and therefore nothing is either generated or destroyed, since this sort of entity is always conserved.[401] [An erroneous hypothesis, but noteworthy as possibly the first naturalistic, and thus testable, explanation for the origin of the world, eschewing supernatural and mythological causes.]

THALES OF MILETUS (C.624-C.545 BCE) GREEK PHILOSOPHER

If ignorance of nature gave birth to the gods, knowledge of nature will dispatch them.[402]

PAUL-HENRI THIRY (1723-1789) FRENCH PHILOSOPHER

Theology is but ignorance of natural causes reduced to a system.[403]

PAUL-HENRI THIRY (1723-1789) FRENCH PHILOSOPHER

Theology is a continual insult to human reason.[404]

PAUL-HENRI THIRY (1723-1789) FRENCH PHILOSOPHER

The hero of my tale, whom I love with all the power of my soul, whom I have tried to portray in all his beauty, who has been, is, and will be beautiful, is - the truth.[405]

LEO TOLSTOY (1828-1920) RUSSIAN WRITER

It is not error which opposes the progress of truth; it is indolence, obstinacy, the spirit of routine, every thing which favors inaction.[406]

ANNE ROBERT JACQUES TURGOT (1727-1781) FRENCH STATESWOMAN

I am not very impressed with theological arguments whatever they may be used to support. Such arguments have often been found unsatisfactory in the past.[407] [An exemplar of understatement.]

ALAN TURING (1912-1954) BRITISH MATHEMATICIAN

Supposing is good, but finding out is better.[408]

MARK TWAIN (1835-1910) AMERICAN WRITER

Faith is believing what you know ain't so.[409]

MARK TWAIN (1835-1910) AMERICAN WRITER

The good thing about science is that it's true whether or not you believe it.[410]

NEIL DEGRASSE TYSON (1958-) AMERICAN ASTROPHYSICIST

Science literacy is not only about what you know, it's about your depth of curiosity, and your capacity to evaluate evidence.[411]

NEIL DEGRASSE TYSON (1958-) AMERICAN ASTROPHYSICIST

Each discovery of science adds a rung to a ladder of knowledge whose beginning is in plain view but whose end is not in sight - we are building the ladder as we go along. By any measure of the ladder's height, we are closer to the beginning of science than to its end.[412]

NEIL DEGRASSE TYSON (1958-) AMERICAN ASTROPHYSICIST

The Arc of Enlightenment in a rational, civilized world: Data > Facts > Information > Knowledge > Understanding > Wisdom.[413]

NEIL DEGRASSE TYSON (1958-) AMERICAN ASTROPHYSICIST

It is a remarkable illustration of the ranging power of the human intellect that a principle first detected in connection with the clumsy puffing of the early steam engines should be found to apply to the whole world, and possibly, even to the whole cosmic universe.[414] [A reference to the Second Law of Thermodynamics.]

ALFRED UBBELOHDE (1907-1988) BELGIAN-ENGLISH CHEMIST

The martyr: I'd rather die than change my mind.[415]

PAUL VALÉRY (1871-1945) FRENCH POET

Science, my boy, is composed of errors, but errors that are correct to make, for they lead step by step to the truth.[416]

JULES VERNE (1828-1905) FRENCH WRITER

He who thinks little, errs much.[417]

LEONARDO DA VINCI (1452-1519) ITALIAN POLYMATH

There is no certainty where one can neither apply any of the mathematical sciences nor any of those which are based upon the mathematical sciences.[418]

LEONARDO DA VINCI (1452-1519) ITALIAN POLYMATH

To burn a book of reasoning is to say, we do not have enough wit to reply to it.[419]

VOLTAIRE (1694-1778) FRENCH PHILOSOPHER

It is obvious the nose has been formed to bear spectacles - thus we have spectacles.[420] [An early, astute pronouncement on the rigor of creationist dogma?]

VOLTAIRE (1694-1778) FRENCH PHILOSOPHER

Let us work without reasoning; it is the only way to make life endurable.[421] [Voiced sardonically.]

VOLTAIRE (1694–1778) FRENCH PHILOSOPHER

Species do not vary continuously in one direction, but vary continuously in many directions; thus there is always material for natural selection to act upon in some direction that may be advantageous.[422] [Affirming that natural selection is not a goal-directed process.]

ALFRED RUSSELL WALLACE (1823–1913) BRITISH NATURALIST

The surest sign that intelligent life exists elsewhere in the universe is that none of it has tried to contact us.[423]

BILL WATTERSON (1958–) AMERICAN CARTOONIST

Heresies are experiments in man's unsatisfied search for truth.[424]

H. G. WELLS (1866–1946) BRITISH WRITER

Giving up witchcraft is, in effect, giving up the Bible.[425] [Yes, John, it is.]

JOHN WESLEY (1703–1791) ENGLISH THEOLOGIAN

Can we ever expect to understand existence? Surely someday we will grasp the central idea of it all as so simple, so beautiful, so compelling that we will all say to each other, "Oh, how could it have been otherwise! How could we all have been so blind so long!"[426]

JOHN ARCHIBALD WHEELER (1911–2008) AMERICAN PHYSICIST

Hypotheses ought to foretell phenomena which have not yet been observed.[427] [Affirming that consequential hypotheses must predict novel phenomena, not merely explain existing phenomena.]

WILLIAM WHEWELL (1794–1866) ENGLISH POLYMATH

We need very much a name to describe a cultivator of science in general. I should incline to call him a Scientist.[428] [From the inventor of the word "scientist".]

WILLIAM WHEWELL (1794–1866) ENGLISH POLYMATH

Aristotle discovered all the half-truths which were necessary to the creation of science.[429] [Acknowledging the importance of first attempts on the long journey toward truth.]

ALFRED NORTH WHITEHEAD (1861–1947) ENGLISH PHILOSOPHER

The candor of science is the glory of the modern. It does not hide and repress; it confronts, turns on the light.[430]

WALT WHITMAN (1819–1892) AMERICAN WRITER

For what is truth? In matters of religion, it is simply the opinion that has survived.[431]

OSCAR WILDE (1854–1900) IRISH PLAYWRIGHT

Tis certain that our senses are extremely disproportioned for comprehending the whole compass and latitude of things.[432] [A pre-Enlightenment point of view which, as was the custom, greatly underestimated the potential of reason and science.]

JOHN WILKINS (1614–1672) ENGLISH THEOLOGIAN

The whole attitude to doubt is, indeed, a critical test of the difference between faith and scientific knowledge.[433]

BARBARA WOOTTON (1897–1988) BRITISH SOCIOLOGIST

Science asks no questions about the ontological pedigree or a priori character of a theory, but is content to judge it by its performance; and it is thus that knowledge ... maintains a strict neutrality toward all philosophical systems and concerns itself not at all with the genesis or a priori grounds of ideas.[434]

CHAUNCEY WRIGHT (1830-1875) AMERICAN PHILOSOPHER

I once wanted to become an atheist, but I gave up. They have no holidays.[435]

HENNY YOUNGMAN (1906-1988) AMERICAN COMEDIAN

MISCELLANY

No exact date marks the beginning of the Enlightenment. Many historians cite 1715, the year that brought an end to Louis XIV's absolutist rule of France. Other scholars date the period from the publication of a major Enlightenment text. Here, Isaac Newton's *Principia Mathematica* is commonly cited, giving a date of 1687. Still others reach further back in time, citing Galileo Galilei's *Dialogue Concerning the Two Chief World Systems* (1632) or Nicolaus Copernicus' *On the Revolutions of the Celestial Spheres* (1543).

ON ETHICS

A man must be excessively stupid, as well as uncharitable, who believes there is no virtue but on his own side.[436]

JOSEPH ADDISON (1672–1719) ENGLISH WRITER

God [is but] a reflection of man's concerns.[437]

ALFRED ADLER (1870–1937) AUSTRIAN PSYCHOTHERAPIST

A wrong life cannot be rightly lived.[438]
THEODOR ADORNO (1903–1969) GERMAN SOCIOLOGIST

The almost insoluble task is to let neither the power of others, nor our own powerlessness, enfeeble us.[439]
THEODOR ADORNO (1903–1969) GERMAN SOCIOLOGIST

There is scarcely a page in the Bible on which an open mind does not perceive a contradiction, an unlikely story, an obvious error, an historical impossibility of one sort or another, so that the intelligence finally is no longer able to accommodate the absurd prejudice that the Bible is totally without error.[440]
STEVE ALLEN (1921–2000) AMERICAN WRITER

Not only is there no God, but try getting a plumber on weekends.[441]
WOODY ALLEN (1935–) AMERICAN FILMMAKER

We are always making God our accomplice, [so that] we may legalize our own iniquities.[442]
HENRI-FRÉDÉRIC AMIEL (1821–1881) SWISS PHILOSOPHER

The Bible addresses him with many names: Satan, Lucifer, Beelzebul ... and so on. Perhaps the most precise name, the one that fits him best, is blasphemy.[443] [A standard-issue shibboleth, equating any critique of dogma to the devil.]
GABRIELE AMORTH (1925–2016) EXORCIST OF THE DIOCESE OF ROME

The needs of a society determine its ethics.[444]
MAYA ANGELOU (1928–2014) AMERICAN WRITER

I distrust those people who know so well what God wants them to do, because I notice it always coincides with their own de-

sires.[445]

SUSAN B. ANTHONY (1820-1906) AMERICAN ACTIVIST

Tell my daughter her father died for freedom of thought and enlightenment principles. And tell her I love her.[446] [A message sent to his eight year-old daughter from prison, where the author awaited a death sentence for insulting Islam.]

SOHEIL ARABI (1985-) IRANIAN BLOGGER

Religious people often prefer to be right rather than compassionate.[447]

KAREN ARMSTRONG (1944-) BRITISH WRITER

[The tyrant] should appear to be particularly earnest in the service of the Gods; for if men think that a ruler is religious and has a reverence for the Gods, they are less afraid of suffering injustice at his hands, and they are less disposed to conspire against him, because they believe him to have the very Gods fighting on his side. At the same time his religion must not be thought foolish.[448] [The philosopher's advice to budding dictators.]

ARISTOTLE (384-322 BCE) GREEK PHILOSOPHER

Never let us deny to this story power and pathos, or treat with hostility ideas which have entered so deep into the life of Christendom. But the story is not true; it never really happened.[449] [Commenting on the Christian promise of salvation.]

MATTHEW ARNOLD (1822-1888) ENGLISH POET

Pious fiction is still fiction.[450]

MATTHEW ARNOLD (1822-1888) ENGLISH POET

Violence is the last refuge of the incompetent.[451] [... and of the disreputable.]

ISAAC ASIMOV (1920–1992) AMERICAN WRITER

They [creationists] always mean beautiful things like hummingbirds. I always reply by saying that I think of a little child in east Africa with a worm burrowing through his eyeball. The worm cannot live in any other way, except by burrowing through eyeballs.[452] [In response to creationists who ask why he does not credit a deity in his nature documentaries.]

DAVID ATTENBOROUGH (1926–) ENGLISH NATURALIST

No species has ever had such wholesale control over everything on earth, living or dead, as we now have. That lays upon us, whether we like it or not, an awesome responsibility.[453]

DAVID ATTENBOROUGH (1926–) ENGLISH NATURALIST

The fact is Christians have long persecuted other sects and each other. People were saying things like, "a woman's place is in the home." And I got to thinking, well, how would someone enforce thoughts like that?[454]

MARGARET ATWOOD (1939–) CANADIAN WRITER

Remember that all is opinion.[455]

MARCUS AURELIUS (121–180 CE) ROMAN EMPEROR

No moral system can rest solely on authority.[456]

A. J. AYER (1910–1989) BRITISH PHILOSOPHER

The greatest vicissitude of things amongst men is the vicissitude of sects and religions.[457]

FRANCIS BACON (1561–1626) ENGLISH NATURAL PHILOSOPHER

The human understanding when it has once adopted an opinion ... draws all things else to support and agree with it. And though there be a greater number and weight of instances to be found on the other side, yet these it either neglects and despises, or ... sets aside and rejects, in order that by this great and pernicious predetermination the authority of its former conclusions may remain inviolate.[458]

FRANCIS BACON (1561–1626) ENGLISH NATURAL PHILOSOPHER

Books must follow sciences, not sciences books.[459]

FRANCIS BACON (1561–1626) ENGLISH NATURAL PHILOSOPHER

One of the many burdens of the person professing Christianity has always been the odium likely to be heaped upon him by fellow Christians quick to smell out, denounce and punish fraud, hypocrisy and general unworthiness among those who assert the faith.[460] [Observing that how Christians mistreat fellow Christians sometimes differs little from how they mistreat non-believers.]

RUSSELL BAKER (1925–2019) AMERICAN JOURNALIST

[As] a jealous lover of human liberty, and deeming it the absolute condition of all that we admire and respect in humanity, I reverse the phrase of Voltaire, and say that if God really existed, it would be necessary to abolish him.[461]

MIKHAIL BAKUNIN (1814–1876) RUSSIAN ACTIVIST

A religion ordinarily imposes on those who have discovered the true faith a spiritual duty of liberating the infidels.[462]

JAMES BALDWIN (1924–1987) AMERICAN WRITER

God has been replaced, as he has all over the West, with respectability and air-conditioning.[463]

AMIRI BARAKA (1934–2014) AMERICAN WRITER

You believe in a book which has talking animals, wizards, witches, demons, sticks turning into snakes, food falling from the sky, people walking on water, and all sorts of magical, absurd, and primitive stories; and you say that I am the one who is mentally ill?[464]

DAN BARKER (1949–) FORMER EVANGELICAL PREACHER

If we admit the existence of realms which science has not conquered, we give to religious quacks and obscurantists domains where only too probably they will house superstition.[465] [A truism evinced by religion's countless "God of the Gaps" stratagems.]

ERNEST WILLIAM BARNES (1874–1953) ENGLISH THEOLOGIAN

If there really is [a] God who created the entire universe with all of its glories, and He decides to deliver a message to humanity, He will not use, as His messenger, a person on cable TV with a bad hairstyle.[466]

DAVE BARRY (1947–) AMERICAN WRITER

During my eighty-seven years I have witnessed a whole succession of technological revolutions. But none of them has done away with the need for character in the individual or the ability to think.[467]

BERNARD BARUCH (1870–1965) AMERICAN PHILANTHROPIST

Emancipate thy mind from the idle fears of superstition and [the] wicked arts of priesthood.[468]

JOHN BASKERVILLE (1707–1775) ENGLISH PRINTER

When plunder becomes a way of life for a group of men in a society, they soon create for themselves a legal system that authorizes it and a moral code that glorifies it.[469]

FRÉDÉRIC BASTIAT (1801–1850) FRENCH ECONOMIST

I have the right to persecute you because I am right and you are wrong, but you have no right to persecute me for the same reason.[470]

JACQUES-BÉNIGNE BOSSUET (1627–1704) FRENCH THEOLOGIAN

When those who conduct ecclesiastical affairs hear anyone speak of toleration, they fancy they hear a most frightful and monstrous opinion.[471]

PIERRE BAYLE (1647–1706) FRENCH PHILOSOPHER

It is pure illusion to think that an opinion which passes down from century to century, from generation to generation, cannot be entirely false.[472]

PIERRE BAYLE (1647–1706) FRENCH PHILOSOPHER

In matters of religion, it is very easy to deceive a man, and very hard to undeceive him.[473]

PIERRE BAYLE (1647–1706) FRENCH PHILOSOPHER

There is no religious behavior without a prior exposure to specific religious ideas.[474]

BENJAMIN BEIT-HALLAHMI (1943–) ISRAELI PSYCHOLOGIST

Intercessory prayer itself had no effect on complication-free recovery [in cardiac bypass patients], but certainty of receiving intercessory prayer was associated with a higher incidence of complications.[475] [The result of a 2005 study measuring the effectiveness of prayer.]

HERBERT BENSON (1935–) AMERICAN CARDIOLOGIST

Men there are who think the cause of truth betrayed by exposing it to doubt.[476]

JEREMY BENTHAM (1748–1832) ENGLISH PHILOSOPHER

It is the greatest happiness of the greatest number that is the measure of right and wrong.[477]

JEREMY BENTHAM (1748–1832) ENGLISH PHILOSOPHER

Show me the man and I'll find you the crime.[478] [An oft-unspoken motto of all despotic regimes.]

LAVRENTIY BERIA (1899–1953) SOVIET SECURITY CHIEF

Truth is the cry of all, but the game of a few.[479]

GEORGE BERKELEY (1685–1753) IRISH PHILOSOPHER

In the end, the law doesn't do justice. People do.[480]

PREET BHARARA (1968–) AMERICAN LAWYER

Whatsoever ye would that men should do to you, do ye even so to them.[481] [A variant of the Golden Rule.]

THE BIBLE

Absurdity: A statement of belief manifestly inconsistent with one's own opinion.[482]

AMBROSE BIERCE (1842–C.1914) AMERICAN SATIRIST

Cynic: A blackguard whose faulty vision sees things as they are, not as they ought to be.[483]

AMBROSE BIERCE (1842–C.1914) AMERICAN SATIRIST

Faith: Belief without evidence in what is told by one who speaks without knowledge.[484]

AMBROSE BIERCE (1842–C.1914) AMERICAN SATIRIST

Scriptures: The sacred books of our holy religion, as distinguished from the false and profane writings on which all other faiths are based.[485]

AMBROSE BIERCE (1842-C.1914) AMERICAN SATIRIST

If the present arrangements of society will not admit of woman's free development, then society must be remodeled, and adapted to the great wants of humanity.[486]

ELIZABETH BLACKWELL (1821-1910) BRITISH PHYSICIAN

In my view all salvation for philosophy may be expected to come from Darwin's theory.[487] [Contending that, one day, human psychology will be explained not from spiritual forces, but from natural, physical mechanisms.]

LUDWIG BOLTZMANN (1844-1906) AUSTRIAN PHYSICIST

Every church is an instrument of bondage.[488]

HYPATIA BRADLAUGH BONNER (1858-1935) BRITISH ACTIVIST

[God] is the celebrity-author of the world's best seller. We have made God into the biggest celebrity of all, to contain our own emptiness.[489]

DANIEL J. BOORSTIN (1914-2004) AMERICAN HISTORIAN

The Vatican is against surrogate mothers. Good thing they didn't have that rule when Jesus was born.[490]

ELAYNE BOOSLER (1952-) AMERICAN COMEDIAN

Religion is cultural. People get it from other people, as they get food preferences, musical tastes, politeness and a dress sense.[491]

PASCAL BOYER (?) FRENCH-AMERICAN ANTHROPOLOGIST

Every child is born into the world an atheist.[492]
CHARLES BRADLAUGH (1833-1891) ENGLISH ACTIVIST

When the Fugitive Slave Law was under discussion in North America, large numbers of clergymen of nearly every denomination were found ready to defend this infamous law.[493] [The law required escaped slaves, even those who had reached free states, to be returned to their owners.]
CHARLES BRADLAUGH (1833-1891) ENGLISH ACTIVIST

Anyone who engages in the practice of psychotherapy confronts every day the devastation wrought by the teachings of religion.[494]
NATHANIEL BRANDEN (1930-2014) CANADIAN-AMERICAN PSYCHOTHERAPIST

The religion of one seems madness to another.[495]
THOMAS BROWNE (1605-1682) ENGLISH WRITER

An intelligent, energetic, educated woman cannot be kept in four walls - even satin-lined, diamond-studded walls - without discovering sooner or later that they are still a prison cell.[496]
PEARL S. BUCK (1892-1973) AMERICAN WRITER

To assert that Christianity communicated to man moral truths previously unknown, argues on the part of the asserter either gross ignorance or else wilful fraud.[497]
HENRY THOMAS BUCKLE (1821-1862) ENGLISH HISTORIAN

God and country stand hand in hand; together they bring nothing but oppression and bloodshed.[498]
LUIS BUÑUEL (1900-1983) SPANISH-MEXICAN FILMMAKER

Those who take refuge behind theological barbed wire fences, quite often wish they could have more freedom of thought, but fear the change to the great ocean of scientific truth as they would a cold bath plunge.[499]

LUTHER BURBANK (1849-1926) AMERICAN BOTANIST

While Europe's eye is fix'd on mighty things,
The fate of empires and the fall of kings;
While quacks of State must each produce his plan,
And even children lisp the Rights of Man;
Amid this mighty fuss let me mention,
The Rights of Woman merit some attention.[500]

ROBERT BURNS (1759-1795) SCOTTISH POET

Prayers are to men as dolls are to children. They are not without use and comfort, but it is not easy to take them very seriously.[501]

SAMUEL BUTLER (1835-1902) ENGLISH WRITER

A clergyman can hardly ever allow himself to look facts fairly in the face. It is his profession to support one side; it is impossible, therefore, for him to make an unbiased examination of the other.[502]

SAMUEL BUTLER (1835-1902) ENGLISH WRITER

You touch on a disheartening truth. People never want to be told anything they do not believe already.[503]

JAMES BRANCH CABELL (1879-1958) AMERICAN WRITER

The trouble with born-again Christians is that they are an even bigger pain the second time around.[504]

HERBERT CAEN (1916-1997) AMERICAN JOURNALIST

Kill them all.[505] [The prior's advice, when it was pointed out to him that the Catholic besiegers of Albi were unable to distinguish between the city's Catholic and non-Catholic - but still Christian - residents.]

CAESARIUS OF HEISTERBACH (C.1180-C.1240) GERMAN PRIOR

A curse from the Lord righteously falls not only on the head of the guilty individual but also on all his lineage.[506] [A faith-based attempt to justify punishment of the innocent.]

JOHN CALVIN (1509-1564) FRENCH THEOLOGIAN

Just think about Irish history, the Middle East, the Crusades, the Inquisition, our own abortion-doctor killings and, yes, the World Trade Center to see how seriously religious people take Thou Shalt Not Kill. Apparently, to religious folks - especially the truly devout - murder is negotiable. It just depends on who's killing and who's getting killed.[507]

GEORGE CARLIN (1937-2008) AMERICAN COMEDIAN

I think it's one of the scars in our culture that we have too high an opinion of ourselves. We align ourselves with the angels instead of the higher primates.[508]

ANGELA CARTER (1940-1992) ENGLISH WRITER

Let no one come to us who has been instructed, or who is wise and prudent, for such qualifications are deemed evil by us.[509] [A satirical assessment of spiritual worthiness.]

CELSUS (CIRCA 150 CE) GREEK PHILOSOPHER

Brutal men with unlimited power are the same all over the world.[510] [From her observations during the American Civil War.]

MARY BOYKIN CHESNUT (1823-1886) AMERICAN WRITER

Anti-black, anti-female, and all forms of discrimination are equivalent to the same thing - anti-humanism.[511]

SHIRLEY CHISHOLM (1924-2005) AMERICAN STATESWOMAN

As our own species is in the process of proving, one cannot have superior science and inferior morals. The combination is unstable and self-destroying.[512]

ARTHUR C. CLARKE (1917-2008) BRITISH WRITER

The predominant emphasis on the motive of fear for the enforcement of absolute commands has made religious morality develop the most intense cruelty that the human heart has known.[513]

MORRIS COHEN (1880-1947) AMERICAN PHILOSOPHER

Whenever philosophy has taken into its plan religion, it has ended in scepticism; and whenever religion excludes philosophy, or the spirit of free inquiry, it leads to wilful blindness and superstition.[514]

SAMUEL TAYLOR COLERIDGE (1772-1834) ENGLISH PHILOSOPHER

There is this difference between the church of Rome and the church of England; the one professes to be infallible, the other to be never in the wrong.[515]

CHARLES COLTON (1780-1832) ENGLISH CLERIC

There are only two things in which the false professors of all religions have agreed: to persecute all other sects and to plunder their own.[516]

CHARLES COLTON (1780-1832) ENGLISH CLERIC

Religion has treated knowledge sometimes as an enemy, sometimes as an hostage, often as a captive, and more often as a child.[517]

CHARLES COLTON (1780-1832) ENGLISH CLERIC

What I do not wish others to do to me, that also I wish not to do to others.[518]

CONFUCIUS (551-479 BCE) CHINESE PHILOSOPHER

Freedom of thought had gradually taken the place in my religion which freedom of the slave occupied before it was secured.[519]

MONCURE CONWAY (1832-1907) AMERICAN ABOLITIONIST

War is an invention of the human mind. The human mind can invent peace with justice.[520]

NORMAN COUSINS (1915-1990) AMERICAN JOURNALIST

What a man really says when he says that someone else can be persuaded by force, is that he himself is incapable of more rational means of communication.[521]

NORMAN COUSINS (1915-1990) AMERICAN JOURNALIST

It's discouraging to think how many people are shocked by honesty and how few by deceit.[522]

NOËL COWARD (1899-1973) ENGLISH PLAYWRIGHT

The people of Northern Ireland are ruled by a regime that is less concerned with morality than with religious doctrine. Their sectarianism is so divisive that, if you say you are an atheist, they ask, "But is it the God of the Catholics or the God of the Protestants in whom you do not believe?"[523]

QUENTIN CRISP (1908-1999) ENGLISH WRITER

Men ... see clearly enough the barbarity of all ages except their own.[524]

ERNEST HOWARD CROSBY (1856-1907) AMERICAN REFORMER

Social instincts lead an animal to take pleasure in the society of its fellows, to feel a certain amount of sympathy with them, and to perform various services for them.[525] [An observation on the evolutionary origins of morality.]

CHARLES DARWIN (1809-1882) ENGLISH NATURALIST

They [blacks] were ... untaught in all the useful arts and occupations, reared in heathen darkness, and, sold by heathen masters, they were transferred to shores enlightened by the rays of Christianity. There, put to servitude, they were trained in the gentle arts of peace and order and civilization; they increased from a few unprofitable savages to millions of efficient Christian laborers.[526] [Smack palm to forehead.]

JEFFERSON DAVIS (1808-1889) CONFEDERATE PRESIDENT

School systems teach the children that they are nothing but glorified apes who have evolutionized out of some primordial soup of mud.[527] [A common attack on the teaching of evolution from a born-again Christian.]

TOM DELAY (1947-) AMERICAN CONGRESSMAN

Can any rational person believe that the Bible is anything but a human document? We now know pretty well where the various books came from, and about when they were written. We know that they were written by human beings who had no knowledge of science, little knowledge of life, and were influenced by the barbarous morality of primitive times, and were grossly ignorant of things that men know today.[528]

CLARENCE DARROW (1857-1938) AMERICAN LAWYER

When the missionaries arrived they fell on their knees and prayed. Then they got up, fell on the Indians, and preyed.[529]

VINE DELORIA, JR. (1933-2005) NATIVE AMERICAN WRITER

People ache to believe that we human beings are vastly different from all other species.[530]

DANIEL DENNETT (1942-) AMERICAN COGNITIVE SCIENTIST

Theologian: The only kind of scholar who has no knowledge of his purported object of study.[531]

KARLHEINZ DESCHNER (1924-2014) GERMAN WRITER

A god that rewards credulity ... is unworthy of any kind of reverence or respect or devotion.[532]

MATT DILLAHUNTY (1969-) AMERICAN ACTIVIST

Eskimo: *If I did not know about God and sin, would I go to hell?*
Priest: *No, not if you did not know.*
Eskimo: *Then why did you tell me?*[533]

ANNIE DILLARD (1945-) AMERICAN WRITER

The myths which are related to Hades, in spite of the fact that their subject matter is fictitious, contribute greatly to fostering piety.[534] [To Hades, or to any other god.]

DIODORUS OF SICILY (CIRCA 50 BCE) GREEK HISTORIAN

No evidence is powerful enough to force acceptance of a conclusion that is emotionally distasteful.[535]

THEODOSIUS DOBZHANSKY (1900-1975) RUSSIAN-AMERICAN BIOLOGIST

The race of men, while sheep in credulity, are wolves for conformity.[536]

CARL VAN DOREN (1885-1950) AMERICAN WRITER

We have men sold to build churches, women sold to support the gospel, and babes sold to purchase Bibles for the poor heathen! All for the glory of God and the good of souls ! The slave auctioneer's bell and the church-going bell chime in with each other, and the bitter cries of the heart-broken slave are drowned in the religious shouts of his pious master.[537]

FREDERICK DOUGLASS (1818–1895) AMERICAN ABOLITIONIST

Until she spoke, no Christian nation had abolished negro slavery. Until she spoke, no Christian nation had given to the world an organized effort to abolish slavery. Until she spoke, the slave trade was sanctioned by all the Christian nations of the world, and our land of liberty and light included. Until Haiti spoke, the church was silent, and the pulpit was dumb.[538] [Reflecting on the successful anti-slavery insurrection in Haiti in 1791.]

FREDERICK DOUGLASS (1818–1895) AMERICAN ABOLITIONIST

Strange how people who suffer together have stronger connections than people who are most content.[539]

BOB DYLAN (1941–) AMERICAN MUSICIAN

We do not want a religion that deceives us for our own good.[540]

ARTHUR EDDINGTON (1882–1944) ENGLISH PHYSICIST

The Bible [is] a collection of venerable but still rather primitive legends.[541]

ALBERT EINSTEIN (1879–1955) GERMAN PHYSICIST

Out here, no one knows where I am, and I am believed to be missing.[542] [Einstein, writing to his sister in 1922 from an undisclosed location in northern Germany, to which he had fled. Police in Berlin, after the murder of a Jewish diplomat by right-wing extremists, warned Einstein that he might be next.]

ALBERT EINSTEIN (1879-1955) GERMAN PHYSICIST

While religion prescribes brotherly love in the relations among ... individuals and groups, the actual spectacle more resembles a battlefield than an orchestra.[543]

ALBERT EINSTEIN (1879-1955) GERMAN PHYSICIST

The man who is thoroughly convinced of the universal operation of the law of causation cannot for a moment entertain the idea of a being who interferes in the course of events.[544]

ALBERT EINSTEIN (1879-1955) GERMAN PHYSICIST

Ethical axioms are found and tested not very differently from the axioms of science. Truth is what stands the test of evidence.[545]

ALBERT EINSTEIN (1879-1955) GERMAN PHYSICIST

[The preacher] has an immense advantage over all other public speakers. The platform orator is subject to the criticism of hisses and groans. Counsel for the plaintiff expects the retort of counsel for the defendant. The honorable gentleman on one side of the House is liable to have his facts and figures shown up by his honorable friend on the opposite side. Even the scientific or literary lecturer, if he is dull or incompetent, may see the best part of his audience slip quietly out one by one. But the preacher is completely master of the situation: no one may hiss, no one may depart. Like the writer of imaginary conversations, he may put what imbecilities he pleases into the mouths of his antagonists, and swell with triumph when he has refuted them. He may riot in gratuitous assertions, confident that no man will contradict him; he may exercise perfect free will in logic, and invent illustrative experience; he may give an evangelical edition

of history with the inconvenient facts omitted - all this he may do with impunity, certain that those of his hearers who are not sympathizing are not listening. For the Press has no band of critics who go the round of the churches and chapels, and are on the watch for a slip or defect in the preacher to make a "feature" in their article: the clergy are, practically, the most irresponsible of all talkers.[546]

GEORGE ELIOT (1819-1880) ENGLISH WRITER

Many good people still believe that support for Christianity is a public duty, however absurd it feels, because other people (though not themselves) cannot be made good without it.[547]

WILLIAM EMPSON (1906-1984) ENGLISH LITERARY CRITIC

If he wants to but cannot, he is weak. If he can but wants not, he is hateful.[548] [On the willingness of a deity, any deity, to prevent evil.]

EPICURUS (341-270 BCE) GREEK PHILOSOPHER

The company of just and righteous men is better than wealth and fertile land.[549]

EURIPIDES (C.480-406 BCE) GREEK PLAYWRIGHT

With the power of government supporting them, at various times and places, Catholics had persecuted Protestants, Protestants had persecuted Catholics, Protestant sects had persecuted other Protestant sects, Catholics of one shade of belief had persecuted Catholics of another shade of belief, and all of these had from time to time persecuted Jews. In efforts to force loyalty to whatever religious group happened to be on top and in league with the government of a particular time and place, men and women had been fined, cast in jail, cruelly tortured, and killed. Among the offenses for which these punishments had been inflicted were such things as speaking disrespectfully of the views of ministers of government-established churches, non-attendance at those churches, expressions of non-belief in their doctrines, and failure to pay taxes and tithes to support them.[550]

EVERSON V BOARD OF EDUCATION (1947) UNITED STATES SUPREME COURT

There exists no politician in India daring enough to attempt to explain to the masses that cows can be eaten.[551] [Commenting, in an interview with Indira Gandhi, on the interplay between religious belief and famine.]

ORIANA FALLACI (1929-2006) ITALIAN JOURNALIST

The true negro does not want integration. He realizes his potential is far better among his own race. We see the hand of Moscow in the background. We see the Devil himself behind it. It will destroy our race eventually.[552] [A Christian expression of brotherhood and love.]

JERRY FALWELL (1933-2007) AMERICAN TELEVANGELIST

Let us begin by stating the conditions under which we would expect to observe increased fervor following the disconfirmation of a belief.[553] [Noting how the unambiguous refutation of a deeply held conviction can strengthen one's belief in the conviction.]

LEON FESTINGER (1919-1989) AMERICAN PSYCHOLOGIST

A man with a conviction is a hard man to change.[554]

LEON FESTINGER (1919-1989) AMERICAN PSYCHOLOGIST

Theology is Anthropology.[555]

LUDWIG FEUERBACH (1804-1872) GERMAN PHILOSOPHER

Man is God.[556] [Asserting that god, any god, is a human projection.]

LUDWIG FEUERBACH (1804–1872) GERMAN PHILOSOPHER

Whenever morality is based on theology, whenever right is grounded on divine authority, the most immoral, unwarranted, infamous things can be justified and established.[557]

LUDWIG FEUERBACH (1804–1872) GERMAN PHILOSOPHER

Man has conceived a God in his own likeness. It is in the name of this pretended God that monarchs and pontiffs have in all the ages and under cover of all religions bound humanity in a slavery from which it has not yet freed itself.[558]

CAMILLE FLAMMARION (1842–1925) FRENCH ASTRONOMER

It was not the multitude, but the sacred specialists, who built up the gigantic and elaborate edifice of theology, which is a purely arbitrary construction, deriving all its design and coherence from the instinctive logic of the human mind, that operates alike in a fairy tale and in a syllogism.[559]

GEORGE WILLIAM FOOTE (1850–1915) ENGLISH SECULARIST

I have no [human] friends. The more you learn about the dignity of the gorilla, the more you want to avoid people.[560]

DIAN FOSSEY (1932–1985) AMERICAN PRIMATOLOGIST

Christianity has done a great deal for love by making it a sin.[561]

ANATOLE FRANCE (1844–1924) FRENCH WRITER

If it were absolutely necessary to choose, I would rather commit an immoral act than a cruel one.[562]

ANATOLE FRANCE (1844–1924) FRENCH WRITER

I think it better to maintain a profound silence with regard to Christian fables: they are canonized by their antiquity and by the credulity of men, each one absurd and puerile for men who reason accordingly.[563]

FREDERICK II (1712–1786) PRUSSIAN KING

Theologians are all alike, of whatever religion or nation they may be; their purpose is always to arrogate to themselves a despotic authority over the conscience of men. They become zealous persecutors of all those who dare to reveal the truth.[564]

FREDERICK II (1712–1786) PRUSSIAN KING

[Religion] comprises a system of wishful illusions together with a denial of reality, such as we find in an isolated form only in the case of amentia, a state of blissful hallucinatory confusion.[565]

SIGMUND FREUD (1856–1939) AUSTRIAN NEUROLOGIST

Religions have never failed to recognize the role of guilt in civilization. They claim to redeem mankind from this sense of guilt, which they call sin.[566]

SIGMUND FREUD (1856–1939) AUSTRIAN NEUROLOGIST

A religion, even if it is called the religion of love, must be hard and unloving towards those who do not belong to it.[567]

SIGMUND FREUD (1856–1939) AUSTRIAN NEUROLOGIST

If I had a large amount of money I should certainly found a hospital for those whose grip upon the world is so tenuous that they can be severely offended by words and phrases and yet remain all unoffended by the injustice, violence and oppression that howls daily about our ears.[568]

STEPHEN FRY (1957–) BRITISH WRITER

I am a lover of truth, a worshipper of freedom, a celebrant at the altar of language and purity and tolerance. That is my religion.[569]

STEPHEN FRY (1957-) BRITISH WRITER

The careful student of history will discover that Christianity has been of very little value in advancing civilization, but has done a great deal toward retarding it.[570]

MATILDA JOSLYN GAGE (1826-1898) AMERICAN SUFFRAGIST

The Parliament of Toulouse, [charged with operating the Catholic Inquisition,] burned 400 witches at one time. Four hundred women at one hour on the public square, dying the horrid death of fire for a crime which never existed save in the imagination of those persecutors and which grew in their imagination from a false belief in woman's extraordinary wickedness, based upon a false theory as to original sin.[571]

MATILDA JOSLYN GAGE (1826-1898) AMERICAN SUFFRAGIST

I, Galileo Galilei, kneeling before your Most Eminent and Most Reverend Lords Cardinals, abjure the false opinion which maintains that the sun is the centre of the universe and is immovable, and that the earth is not the centre and is movable, after it hath been signified to me that the said doctrine is repugnant with the Holy Scripture.[572] [Galileo's forced recantation of the theory of heliocentrism.]

GALILEO GALILEI (1564-1642) ITALIAN PHYSICIST

An error does not become truth by reason of multiplied propagation.[573]

MOHANDAS GANDHI (1869-1948) INDIAN ACTIVIST

With reasonable men, I will reason; with humane men I will plead; but to tyrants I will give no quarter, nor waste arguments where they will certainly be lost.[574]

WILLIAM LLOYD GARRISON (1805-1879) AMERICAN ABOLITIONIST

Five hallmarks of pseudoscientists: They argue that the scientific consensus emerges from a conspiracy to suppress dissenting views. They produce fake experts, who have views contrary to established knowledge but do not actually have a credible scientific track record. They cherry-pick the data and papers that challenge the dominant view as a means of discrediting an entire field. They deploy false analogies and other logical fallacies. And they set impossible expectations of research - when scientists produce one level of certainty, the pseudoscientists insist they achieve another.[575]

ATUL GAWANDE (1965-) AMERICAN PHYSICIAN

The delegates of the annual conference ... are decidedly opposed to modern abolitionism, and wholly disclaim any right, wish, or intention to interfere in the civil and political relation between master and slave, as it exists in the slaveholding states of this Union.[576] [One of religion's less illustrious moments.]

GENERAL CONFERENCE RESOLUTION #2 (1836) METHODIST EPISCOPAL CHURCH

Blasphemy: A law to protect an all-powerful, supernatural deity from getting its feelings hurt.[577]

RICKY GERVAIS (1961-) ENGLISH COMEDIAN

Those who send videos of themselves removing the hijab will be sentenced to between one and ten years in prison.[578] [A warning against posting such images on the social media site of Masih Alinejad, an American-Iranian activist who champions women's rights and challenges the forced wearing of headscarves.]

MOUSA GHAZANFARABADI (?-) IRANIAN JURIST

The various modes of worship were all considered by the people, as equally true; by the philosopher, as equally false; and by the magistrate, as equally useful.[579]

EDWARD GIBBON (1737-1794) ENGLISH HISTORIAN

A policy conceived only as the conquest and maintenance of power is the most hostile thing to man's desire.[580]

MARY ANN GLENDON (1938-) AMERICAN SCHOLAR

The church alone,
beyond all question,
has for ill-gotten gains,
the right digestion.[581]

JOHANN WOLFGANG VON GOETHE (1749-1832) GERMAN WRITER

Christianity is most admirably adapted to the training of slaves.[582]

EMMA GOLDMAN (1869-1940) RUSSIAN WRITER

The greatest danger to our future is apathy.[583]

JANE GOODALL (1934-) ENGLISH PRIMATOLOGIST

If one small and odd lineage of fishes had not evolved fins capable of bearing weight on land (though evolved for different reasons in lakes and seas), terrestrial vertebrates would never have arisen. If a large extraterrestrial object - the ultimate random bolt from the blue - had not triggered the extinction of dinosaurs sixty-five million years ago, mammals would still be small creatures, confined to the nooks and crannies of a dinosaur's world, and incapable of evolving the larger size that brains big enough for self-consciousness require. If a small and tenuous population of protohumans had not survived a hundred slings and arrows of outrageous fortune (and potential ex-

tinction) on the savannas of Africa, then Homo sapiens would never have emerged to spread throughout the globe. We are glorious accidents of an unpredictable process with no drive to complexity, not the expected results of evolutionary principles that yearn to produce a creature capable of understanding the mode of its own necessary construction.[584]

STEPHEN J. GOULD (1941-2002) AMERICAN PALEONTOLOGIST

A measure which makes at once four million people voters, who were heretofore declared by the highest tribunal in the land not citizens of the United States ... is indeed a measure of grander importance than any other one act of the kind from the foundation of our free Government to the present time.[585] [Commenting, in 1870, on the passage of the Fifteenth Amendment, which granted the right to vote to all male citizens, regardless of race or color. Women had to wait another fifty years for the same right.]

ULYSSES S. GRANT (1822-1885) AMERICAN PRESIDENT

The gods themselves are on a learning curve.[586] [Commenting on the many inconsistencies in religious doctrines.]

BARBARA GRAZIOSI (?) ITALIAN CLASSICIST

Religious freedom is never about stopping persecution; it's about being the one who gets to do it.[587] [On the veiled ambition of any religion that seeks political power.]

DEVEN GREEN (1975-) CANADIAN COMEDIAN

It is a mark of privilege to live in a society where religion is weakened, where it displays its sweet and kind face. Many Westerners have only known faith in retreat, and do not know the brutality of religion when it has nothing to fear.[588]

SARAH HAIDER (?) PAKISTANI-AMERICAN WRITER

Once you get people believing that there is an authoritative well of wisdom to which they can turn for absolutes, you have dried up the springs on which they must in the end draw.[589]

LEARNED HAND (1872–1961) AMERICAN JURIST

Faith is the license that religious people give one another to keep believing when reasons fail, to keep believing in the absence of evidence.[590]

SAM HARRIS (1967–) AMERICAN NEUROSCIENTIST

If you wake up tomorrow morning thinking that saying a few Latin words over your pancakes is going to turn them into the body of Elvis Presley, you have lost your mind. But if you think more or less the same thing about a cracker and the body of Jesus, you are just a Catholic.[591]

SAM HARRIS (1967–) AMERICAN NEUROSCIENTIST

I will win. Not immediately, but definitely.[592] [Arrested, imprisoned, and tortured for the right of women to drive a car.]

LOUJAIN AL-HATHLOUL (1989–) SAUDI ACTIVIST

It is indeed probable that more harm and misery have been caused by men determined to use coercion to stamp out a moral evil than by men intent on doing evil.[593]

FRIEDRICH HAYEK (1899–1992) AUSTRIAN-BRITISH PHILOSOPHER

To preserve any faith by the maintenance of ignorance is sufficient indictment of the faith itself and of those who thus secure its permanence.[594]

ELLEN HAYES (1851–1930) AMERICAN MATHEMATICIAN

Men rarely (if ever) manage to dream up a god superior to themselves. Most gods have the manners and morals of a spoiled child.[595]

ROBERT HEINLEIN (1907–1988) AMERICAN WRITER

Fashions in sin change.[596]

LILLIAN HELLMAN (1905–1984) AMERICAN PLAYWRIGHT

Every judgment teeters on the brink of error. To claim absolute knowledge is to become monstrous. Knowledge is an unending adventure at the edge of uncertainty.[597]

FRANK HERBERT (1920–1986) AMERICAN WRITER

Religion can bear no jesting.[598]

GEORGE HERBERT (1593–1622) WELSH POET

Religions exist primarily to solve human problems.[599]

A. L. HERMAN (1930–) AMERICAN SCHOLAR

All human beings are equal, but all cultures and religions are not. A culture that celebrates femininity and considers women to be the masters of their own lives is better than a culture that mutilates girls' genitals and confines them behind walls and veils or flogs or stones them for falling in love.[600]

AYAAN HIRSI ALI (1969–) DUTCH-AMERICAN ACTIVIST

There is a huge difference between being tolerant and tolerating intolerance.[601]

AYAAN HIRSI ALI (1969–) DUTCH-AMERICAN ACTIVIST

Human decency is not derived from religion. It precedes it.[602]

CHRISTOPHER HITCHENS (1949–2011) ENGLISH-AMERICAN WRITER

Interviewer: *Millions of people died in the Soviet experiment. If you had known that, would it have made a difference to you, to*

your commitment to being a communist?
Hobsbawm: *Probably not.*
Interviewer: So, had the [ideal society] actually been created, the loss of fifteen, twenty million people might have been justified?
Hobsbawm: *Yes.*[603]
[A superb articulation of the zeal, present in all absolutist ideologies, religious or secular, to destroy millions in pursuit of their chimerical nirvanas.]
ERIC HOBSBAWM (1917-2012) BRITISH MARXIST HISTORIAN

There are similarities between absolute power and absolute faith: a demand for absolute obedience; a readiness to attempt the impossible; a bias for simple solutions - to cut the knot rather than unravel it; the viewing of compromise as surrender; the tendency to manipulate people and "experiment with blood".[604]
ERIC HOFFER (1922-1983) AMERICAN PHILOSOPHER

The savior who wants to turn men into angels is as much a hater of human nature as the totalitarian despot who wants to turn them into puppets.[605]
ERIC HOFFER (1922-1983) AMERICAN PHILOSOPHER

There is in us a tendency to judge a race, a nation, or an organization by its least worthy members.[606]
ERIC HOFFER (1922-1983) AMERICAN PHILOSOPHER

Irrationality is the square root of all evil.[607]
DOUGLAS HOFSTADTER (1945-) AMERICAN SCIENTIST

Certitude is not the test of certainty.[608]
OLIVER WENDELL HOLMES, JR. (1841-1935) AMERICAN JURIST

Deep-seated preferences cannot be argued about - you cannot argue a man into liking a glass of beer - and therefore, when differences are sufficiently far reaching, we try to kill the other man rather than let him have his way. But that is perfectly consistent with admitting that, so far as appears, his grounds are just as good as ours.[609]

OLIVER WENDELL HOLMES, JR. (1841-1935) AMERICAN JURIST

The historical evidence reinforces this conclusion: The history of religious struggle and conflicts has not been primarily an attempt to increase the sphere of political and personal freedom, but rather to exercise dominion with respect to religious dogma or practice.[610]

SIDNEY HOOK (1902-1989) AMERICAN PHILOSOPHER

Religious tolerance has developed more as a consequence of the impotence of religions to impose their dogmas on each other than as a consequence of spiritual humility.[611]

SIDNEY HOOK (1902-1989) AMERICAN PHILOSOPHER

The men trained to it [theology] can do nothing else, and so they clutch it and cling to it, declaring it the hope of the world, and [they] prophecy terrible things that will happen to a godless government.[612]

ELBERT HUBBARD (1856-1915) AMERICAN WRITER

At the base, the people; above the people, religion represented by the clergy; by the side of religion, justice represented by the magistracy. And, at that period of human society, what was the people? It was ignorance. What was religion? It was intolerance. And what was justice? It was injustice.[613] [Commenting on the structure of French society before the Revolution.]

VICTOR HUGO (1802-1885) FRENCH NOVELIST

The heroes of paganism correspond exactly to the saints in popery and Islam. The place of Hercules, Theseus, Hector, and Romulus is now supplied by Dominic, Francis, Anthony, and Benedict.[614]

DAVID HUME (1711–1776) BRITISH PHILOSOPHER

The sense of justice and injustice is not derived from nature, but arises from education and human conventions.[615]

DAVID HUME (1711–1776) BRITISH PHILOSOPHER

Seek the truth, hear the truth, learn the truth, love the truth, speak the truth, hold the truth and defend the truth until death.[616] [Burned at the stake for challenging Holy Writ.]

JAN HUS (C.1369–1415) CZECH THEOLOGIAN

A fanatic is a man who consciously over-compensates a secret doubt.[617]

ALDOUS HUXLEY (1894–1963) BRITISH WRITER

[Humanism] will have nothing to do with absolutes, including absolute truth, absolute morality, absolute perfection and absolute authority.[618]

JULIAN HUXLEY (1887–1975) ENGLISH BIOLOGIST

Irrationally held truths may be more harmful than reasoned errors.[619]

THOMAS HUXLEY (1825–1895) ENGLISH BIOLOGIST

I propose to raise a revolution against the lie that the majority has the monopoly of the truth.[620]

HENRIK IBSEN (1828–1906) NORWEGIAN PLAYWRIGHT

The Church hates a thinker precisely for the same reason a rob-

ber dislikes a sheriff, or a thief despises the prosecuting witness.[621]

ROBERT GREEN INGERSOLL (1833-1899) AMERICAN LAWYER

No man of any humor ever founded a religion - never.[622]

ROBERT GREEN INGERSOLL (1833-1899) AMERICAN LAWYER

The clergy know, that I know, that they know, that they do not know.[623]

ROBERT GREEN INGERSOLL (1833-1899) AMERICAN LAWYER

If all the historic books of the Bible were blotted from the memory of mankind, nothing of value would be lost.[624]

ROBERT GREEN INGERSOLL (1833-1899) AMERICAN LAWYER

For spreading corruption and prostitution by taking off her hijab and walking without a veil.[625] [Citing the basis for the 15-year sentence of Iranian civil rights activist Saba Kord Afshari.]

ISLAMIC REVOLUTIONARY COURT (2019)

For assembly and collusion with intention to commit a crime against national security; for propaganda against the state; for encouraging corruption and prostitution; for appearing in public without the hijab; for publishing falsehoods with the intention to disturb public opinion; and for disturbing public order.[626] [Citing the basis for the 33-year, 148-lash sentence of Iranian human rights lawyer Nasrin Sotoudeh.]

ISLAMIC REVOLUTIONARY COURT (2019)

Virtually all ideologues, of any variety, are fearful and insecure, which is why they are drawn to ideologies that promise prefabricated answers for all circumstances.[627]

JANE JACOBS (1916-2006) CANADIAN WRITER

Religion, in short, is a monumental chapter in the history of human egotism.[628]

WILLIAM JAMES (1842–1910) AMERICAN PHILOSOPHER

Je suis Charlie.[629] [A motto of solidarity, after extremists murdered twelve journalists for slighting their religion.]

CHARLIE HEBDO (2015) FRENCH NEWSPAPER

The way to silence religious disputes is to take no notice of them.[630]

THOMAS JEFFERSON (1743–1826) AMERICAN PRESIDENT

The Newtonian principle of gravitation is now more firmly established, on the basis of reason, than it would be were the government to step in, and to make it an article of faith. Reason and experiment have been indulged, and error has fled before them. It is error alone which needs the support of government. Truth can stand by itself.[631]

THOMAS JEFFERSON (1743–1826) AMERICAN PRESIDENT

It does me no injury for my neighbor to say there are twenty gods or no God. It neither picks my pocket nor breaks my leg.[632] [An early rendition of "your rights stop at my nose."]

THOMAS JEFFERSON (1743–1826) AMERICAN PRESIDENT

No man ever had a distinct idea of the trinity. It is the mere Abracadabra of the mountebanks calling themselves the priests of Jesus. If it could be understood it would not answer their purpose, [for] their security is in their faculty of shedding darkness.[633]

THOMAS JEFFERSON (1743–1826) AMERICAN PRESIDENT

What is terrorism? Terrorism in some sense is a reaction

against the creation of a type one civilization. Now most terrorists cannot articulate this. ... What they're reacting to is the fact that we're headed toward a multicultural, tolerant, scientific society, and that is what they don't want. They don't want science. They want a theocracy. They don't want multiculturalism. They want monoculturalism. So instinctively, they don't like the march toward a type one civilization.[634] [On the Kardashev scale, "type one" refers to a planet-wide civilization that is culturally and technologically advanced.]

MICHIO KAKU (1947-) AMERICAN PHYSICIST

The death of dogma is the birth of morality.[635]

IMMANUEL KANT (1724-1804) GERMAN PHILOSOPHER

I ought never to act except in such a way that I could also will that my maxim should become a universal law.[636]

IMMANUEL KANT (1724-1804) GERMAN PHILOSOPHER

A man dies when he refuses to stand up for that which is right. A man dies when he refuses to stand up for justice. A man dies when he refuses to take a stand for that which is true.[637]

MARTIN LUTHER KING (1929-1968) AMERICAN ACTIVIST

If there is a universal mind, must it be sane?[638]

DAMON KNIGHT (1922-2002) AMERICAN WRITER

I am not out to destroy the Christian convictions of people in whom they are deeply implanted and to whom they mean a great deal. And I am sure that nothing I say here will have the slightest effect on believers of this type. But what I do want to argue is that ... it is a mistake to try to impose them on children, and to make them the basis of moral training. The moral education of children is much too important a matter to be built on such foundations.[639]

MARGARET K. KNIGHT (1903–1983) ENGLISH PSYCHOLOGIST

Offenses and public insults committed within Italian territory against the person of the Supreme Pontiff, whether by speech, deed, or writing, will be punished in the same manner as offenses and insults against the person of the King.[640] [One of several protections negotiated by the Catholic Church with the then-fascist Italian state to prevent criticism of the pope. The treaty also provided a financial settlement to the Church, but offered no protections for the laity.]

LATERAN TREATY (1929)

It takes as much time and trouble to pull down a falsehood as to build up a truth.[641]

PETER MERE LATHAM (1789–1875) ENGLISH PHYSICIAN

Give me your tired, your poor,
Your huddled masses yearning to breathe free.
The wretched refuse of your teeming shore.
Send these, the homeless, tempest-tost to me,
I lift my lamp beside the golden door.[642]

EMMA LAZARUS (1849–1887) AMERICAN POET

Almost all Europe, for many centuries, was inundated in blood, which was shed at the direct instigation or with the full approval of the ecclesiastical authorities.[643]

W. E. H. LECKY (1838–1903) IRISH HISTORIAN

That the Church of Rome has shed more innocent blood than any other institution that has ever existed among mankind, will be questioned by no Protestant who has a competent knowledge of history.[644]

W. E. H. LECKY (1838-1903) IRISH HISTORIAN

Imagine there's no countries,
It isn't hard to do,
Nothing to kill or die for,
And no religion, too.[645]

JOHN LENNON (1940-1980) ENGLISH MUSICIAN

A heretic is a man who sees with his own eyes.[646]

GOTTHOLD LESSING (1729-1781) GERMAN PHILOSOPHER

The leaders of organised institutions of religion know that that which is impressed upon the mind of a child before the age of seven, molds and shapes the child's character. Only by the most heroic mental effort are men and women able to eradicate the teachings received in childhood.[647]

JOSEPH LEWIS (1889-1968) AMERICAN ACTIVIST

Religious instruction narrows a child's mentality to the scope of the particular religious dogma it receives. The teaching of ethics and morality, free from religious bias, permits a child to weigh the opinions and thoughts of the world.[648]

JOSEPH LEWIS (1889-1968) AMERICAN ACTIVIST

If slavery is not wrong, nothing in wrong.[649]

ABRAHAM LINCOLN (1809-1865) AMERICAN PRESIDENT

The churches used to win their arguments against atheism, agnosticism, and other burning issues by burning the ism-ists, which is fine proof that there is a devil but hardly evidence that there is a God.[650]

BEN LINDSEY (1869-1943) AMERICAN JURIST

The teachers of a theistic morality, when the audience is devout, have only to fortify the impression that the rules of conduct are certified by God the invisible King. ... But the teachers of humanism have no credentials. Their teaching is not certified. They have to prove their case by the test of mundane experience.[651]

WALTER LIPPMANN (1889-1974) AMERICAN WRITER

No one ought to harm another in his life, health, liberty or possessions.[652]

JOHN LOCKE (1632-1704) ENGLISH PHILOSOPHER

New opinions are always suspected, and usually opposed, without any other reason but because they are not already common.[653]

JOHN LOCKE (1632-1704) ENGLISH PHILOSOPHER

Papists are not to enjoy the benefit of toleration because, where they have power, they think themselves bound to deny it to others.[654]

JOHN LOCKE (1632-1704) ENGLISH PHILOSOPHER

Man appears to be the missing link between anthropoid apes and human beings.[655]

KONRAD LORENZ (1903-1989) AUSTRIAN ZOOLOGIST

Again and again our foe, religion, has given birth to deeds sinful and unholy.[656]

LUCRETIUS (99-55 BCE) ROMAN PHILOSOPHER

God's work and word stare us in the face, declaring that women must be used either for marriage or for fornication.[657] [A viewpoint, not uncommon within religions past and present, on the ethical treatment of women.]

MARTIN LUTHER (1483–1546) GERMAN THEOLOGIAN

Let every person be subject to the governing authorities with fear and reverence.[658] [An admonition to obey civil authorities, even when they are unjust.]

MARTIN LUTHER (1483–1546) GERMAN THEOLOGIAN

It is impossible to deny that the polity of the Church of Rome is the very masterpiece of human wisdom. ... The experience of twelve hundred eventful years, the ingenuity and patient care of forty generations of statesmen, have improved that polity to such perfection that, among the contrivances which have been devised for deceiving and oppressing mankind, it occupies the highest place.[659]

THOMAS BABINGTON MACAULAY (1800–1859) BRITISH HISTORIAN

Christianity was preached by ignorant people but believed by scholars.[660]

JOSEPH DE MAISTRE (1753–1821) FRENCH PHILOSOPHER

Tolerance becomes a crime when applied to evil.[661]

THOMAS MANN (1875–1955) GERMAN WRITER

If a test of civilization be sought, none can be so sure as the condition of that half of society over which the other half has power.[662]

HARRIET MARTINEAU (1802–1876) ENGLISH PHILOSOPHER

Man makes religion, religion does not make man.[663]

KARL MARX (1818–1883) GERMAN PHILOSOPHER

Like all weak men he laid an exaggerated stress on not chan-

ging one's mind.[664]

W. SOMERSET MAUGHAM (1874-1965) ENGLISH WRITER

Man is the ethical animal.[665]

ROLLO MAY (1909-1994) AMERICAN PSYCHOLOGIST

ETHICS

The opposite to courage is not cowardice; it is conformity.[666]

ROLLO MAY (1909-1994) AMERICAN PSYCHOLOGIST

I doubt if France ever did a greater service to its women than when it ... closed their convents.[667]

JOSEPH MCCABE (1867-1955) ENGLISH WRITER

The First Amendment expresses our Nation's fundamental commitment to religious liberty by means of two provisions - one protecting the free exercise of religion, the other barring establishment of religion.[668]

MCCREARY V ACLU (2005) UNITED STATES SUPREME COURT

It is true that many Americans find the Commandments in accord with their personal beliefs. But we do not count heads before enforcing the First Amendment.[669]

MCCREARY V ACLU (2005) UNITED STATES SUPREME COURT

Once you've grasped that the universe is not about you, it sets you free in the most extraordinary way.[670]

IAN MCEWAN (1948-?) ENGLISH NOVELIST

A still tongue makes a happy life.[671] [An axiom of despotic regimes.]

OBSERVED BY THE CHARACTER "NUMBER SIX" (1967) THE PRISONER

There are always those who desire compromise and a middle course ... But where one of the sides in the conflict has the eternal truth, as in the case of those of us who embrace the great doctrines of the historic Christian faith, to compromise in the slightest is to dishonor and destroy the faith.[672] [Here, the quote attacks not secularists or humanists, but fellow Christians who fail to heed the author's fundamentalist interpretation of doctrine.]

CARL MCINTIRE (1906-2002) AMERICAN CLERIC

All forms of violence are quests for identity.[673]

MARSHALL MCLUHAN (1911-1980) CANADIAN EDUCATOR

It was morality that burned the books of the ancient sages, and morality that halted the free inquiry of the Golden Age and substituted for it the credulous imbecility of the Age of Faith. It was a fixed moral code and a fixed theology which robbed the human race of a thousand years by wasting them upon alchemy, heretic-burning, witchcraft and sacerdotalism.[674]

H. L. MENCKEN (1880-1956) AMERICAN JOURNALIST

Morality is the theory that every human act must be either right or wrong, and that 99% of them are wrong.[675]

H. L. MENCKEN (1880-1956) AMERICAN JOURNALIST

The most costly of all fallacies is to believe passionately in the palpably not true.[676]

H. L. MENCKEN (1880-1956) AMERICAN JOURNALIST

Puritanism: The haunting fear that someone, somewhere may be happy.[677]

H. L. MENCKEN (1880-1956) AMERICAN JOURNALIST

Moral certainty is always a sign of cultural inferiority. The more uncivilized the man, the surer he is that he knows precisely what is right and what is wrong.[678]

H. L. MENCKEN (1880-1956) AMERICAN JOURNALIST

Say what you will about the ten commandments, you must always come back to the pleasant fact that there are only ten of them.[679]

H. L. MENCKEN (1880-1956) AMERICAN JOURNALIST

Tolerance is an expression of respect for other people's history, traditions, religion and identities. But let no one be mistaken; tolerance does not mean "anything goes". There must be zero tolerance towards all who violate and trample on the inalienable rights of man.[680]

ANGELA MERKEL (1954-) GERMAN CHANCELLOR

How I suffered when I had to preach to you those pious lies that I detested in my heart. What contempt I had for my ministry. A thousand times I was on the brink of declaring so publicly, thus opening your eyes, but fear constrained me, and forced me to keep silence until my death.[681] [Meslier's confession, disclosed posthumously.]

JEAN MESLIER (1664-1729) FRENCH PRIEST

Saint Jerome explicitly says in many places of his Prologues that the texts had been corrupted and falsified, having already in his time been through the hands of all kinds of persons who added and cut whatever they liked, with the result, he said, that there were as many different readings as there were different texts.[682] [Commenting on the accuracy of scripture.]

JEAN MESLIER (1664-1729) FRENCH PRIEST

If atheism ... were generally spread, all branches of religion would be ... cut off by the root. No more theological wars.[683]

JULIEN OFFRAY DE LA METTRIE (1709-1751) FRENCH PHYSICIAN

You are still half savage, but there is hope.[684] [An advanced alien, commenting on the morality of *Homo sapiens*.]

VOICED BY THE CHARACTER METRON (1967) STAR TREK

A large proportion of the noblest and most valuable moral teaching has been the work, not only of men who did not know, but of men who knew and rejected, the Christian faith.[685]

JOHN STUART MILL (1806-1873) BRITISH PHILOSOPHER

Bad men need nothing more to compass their ends, than that good men should look on and do nothing.[686]

JOHN STUART MILL (1806-1873) BRITISH PHILOSOPHER

As a social and as a personal force, religion has become a dependent variable. It does not originate; it reacts. It does not denounce; it adapts. It does not set forth new models of conduct and sensibility; it imitates.[687]

C. WRIGHT MILLS (1916-1962) AMERICAN SOCIOLOGIST

When people were committed to the idea that in the field of religion only one plan must be adopted, bloody wars resulted. With the acknowledgment of the principle of religious freedom these wars ceased.[688]

LUDWIG VON MISES (1881–1973) AUSTRIAN ECONOMIST

In outer space you develop an instant global consciousness, a people orientation, an intense dissatisfaction with the state of the world, and a compulsion to do something about it. From out there on the moon, international politics look so petty. You want to grab a politician by the scruff of the neck and drag him a quarter of a million miles out and say, "Look at that, you son of a bitch!"[689]

EDGAR MITCHELL (1930–2016) AMERICAN ASTRONAUT

Any suggestion that Christianity ... created a civil society and enhanced economic performance ... is sheer nonsense.[690] [Commenting on the role of Christianity after the fall of Rome.]

JOEL MOKYR (1946–) AMERICAN-ISRAELI HISTORIAN

Nothing is so firmly believed as what is least known.[691]

MICHEL DE MONTAIGNE (1533–1592) FRENCH PHILOSOPHER

I can assure you that no kingdom has ever had as many civil wars as the kingdom of Christ.[692]

MONTESQUIEU (1689–1755) FRENCH PHILOSOPHER

Do you think God will punish those who failed to practice a religion that was not revealed to them?[693] [Yes, according to no small number of theologians.]

MONTESQUIEU (1689–1755) FRENCH PHILOSOPHER

I admit that history is full of religious wars, but let one take care to observe: it is not the multiplicity of religions that has produced these wars, it is the intolerant spirit of whichever religion believes itself dominant.[694]

MONTESQUIEU (1689–1755) FRENCH PHILOSOPHER

The admission of slaves into the representation, when fairly explained, comes to this: that the inhabitant of Georgia and South Carolina, who goes to the coast of Africa, and, in defiance of the most sacred laws of humanity, tears away his fellow-creatures from their dearest connections, and damns them to the most cruel bondage, shall have more votes, in a government instituted for the protection of the rights of mankind, than the citizen of Pennsylvania or New Jersey.[695] [Arguing, in one of many such instances, against concessions to slavery in the Constitution of the United States.]

GOUVERNEUR MORRIS (1752–1816) AMERICAN STATESMAN

[Many profess religious belief] not quite because they believe it to be true, but because they are strongly convinced that it is good for other people to believe it.[696]

GILBERT MURRAY (1866–1957) AUSTRALIAN-BRITISH PHILOSOPHER

There is no social evil, no form of injustice, whether of the feudal or capitalist order, which has not been sanctified in some way or another by religious sentiment and thereby rendered more impervious to change.[697]

RHEINHOLD NIEBUHR (1892–1971) AMERICAN THEOLOGIAN

Convictions are more dangerous enemies of truth than lies.[698]

FRIEDRICH NIETZSCHE (1844-C.1900) GERMAN PHILOSOPHER

Is man only a blunder of God, or is God only a blunder of man?[699]

FRIEDRICH NIETZSCHE (1844-C.1900) GERMAN PHILOSOPHER

Perhaps the ant in the forest is as firmly convinced that it is the aim and purpose of the forest, as we are convinced [that we are the aim and purpose] of the world.[700]

FRIEDRICH NIETZSCHE (1844-C.1900) GERMAN PHILOSOPHER

In the beginning there was nonsense, and the nonsense was with God, and the nonsense was God.[701]

FRIEDRICH NIETZSCHE (1844-C.1900) GERMAN PHILOSOPHER

He sewed his eyes shut because he is afraid to see,
He tries to tell me what I put inside of me,
He's got the answers to ease my curiosity,
He dreamed a god up and called it Christianity.[702]

NINE INCH NAILS (1994) AMERICAN MUSICIANS

Religious ideologies and their fanaticisms are dangerous enough, but when these or other ideologies become frenzied elements of the political area, the only area of absolute power over human lives, they become potentially disastrous in their impact upon a free society.[703]

ROBERT NISBET (1913-1996) AMERICAN SOCIOLOGIST

Men have gone to war and cut each other's throats because they could not agree as to what was to become of them after their throats were cut.[704]

NORTH CAROLINA V BEAL (1930) NORTH CAROLINA SUPREME COURT

The loss of religious faith among the most civilized portion of the race is a step from childishness toward maturity.[705]

CHARLES ELIOT NORTON (1827-1908) AMERICAN WRITER

She was a good Christian woman with a large respect for religion, though she did not, of course, believe any of it was true.[706]

FLANNERY O'CONNOR (1925-1964) AMERICAN WRITER

Orgel's Second Rule: *Evolution is a lot cleverer than you are.*[707]

LESLIE ORGEL (1927-2007) BRITISH CHEMIST

The radical defect of Christianity is that it tried to win the world by a bribe, and it has become a nullity.[708]

OUIDA (1839-1908) ENGLISH WRITER

It is revelation to the first person only, and hearsay to every other.[709] [Observing a truth regarding purported communications from the Almighty.]

THOMAS PAINE (1737-1809) ENGLISH ACTIVIST

Twelve centuries of moral and political darkness, in which Europe was involved, had nearly completed the destruction of human dignity, and every thing valuable or ornamental in the character of man. During this long and doleful night of ignorance, slavery and superstition, Christianity reigned triumphant; its doctrines and divinity were not called in question.[710]

ELIHU PALMER (1764-1806) AMERICAN WRITER

You can't teach an old dogma new tricks.[711]

DOROTHY PARKER (1893-1967) AMERICAN WRITER

The arc of the moral universe is long, but it bends toward justice.[712]

THEODORE PARKER (1810-1860) AMERICAN CLERIC

I did not want to be mistreated, I did not want to be deprived of a seat that I had paid for. It was just time.[713] [Recalling her refusal to relinquish her bus seat to a white person which, at the time, the law required.]

ROSA PARKS (1913-2005) AMERICAN ACTIVIST

By the time I was six, I was old enough to realize that we were not free. The Ku Klux Klan was riding through the black community, burning churches, beating up people, killing people.[714]

ROSA PARKS (1913-2005) AMERICAN ACTIVIST

Men believe not from proof but from agreement.[715]

BLAISE PASCAL (1623-1662) FRENCH MATHEMATICIAN

Men never do evil so completely and cheerfully as when they do it from religious conviction.[716]

BLAISE PASCAL (1623-1662) FRENCH MATHEMATICIAN

We must be born guilty, or God would be unjust.[717] [Arguably, one of the strangest justifications for the alleged existence of a caring deity.]

BLAISE PASCAL (1623-1662) FRENCH MATHEMATICIAN

If "humility" means nothing more than the capacity to learn from criticism, then it has an undoubted value; but if "humility" means a willingness to submit to authority - to abandon or to modify what one is doing merely because it does not accord with the teachings of the Bible or the thoughts of Chairman Mao - then it is death to the spirit: the proper name for it, indeed, is "servility".[718]

JOHN PASSMORE (1914-2004) AUSTRALIAN PHILOSOPHER

Never trust a man who reads only one book.[719]

ARTURO PÉREZ-REVERTE (1951-) SPANISH WRITER

It is true, of course, that the Churches remain about the most highly segregated institutions of American life (though country clubs and college fraternities do surpass us).[720]

JAMES A. PIKE (1913-1969) AMERICAN THEOLOGIAN

Knowledge should be his guide, not personal experience.[721]
PLATO (C.428-C.348 BCE) GREEK PHILOSOPHER

The theory of evolution is incapable of establishing the dignity of man.[722] [A pontifical attempt to subsume science under faith.]
POPE JOHN PAUL II (1920-2005)

The attempt to create the Kingdom of Heaven on earth may easily succeed in turning our earth into a hell for our fellow men.[723]
KARL POPPER (1902-1994) AUSTRIAN-BRITISH PHILOSOPHER

If we are not prepared to defend a tolerant society against the onslaught of the intolerant, then the tolerant will be destroyed, and tolerance with them.[724]
KARL POPPER (1902-1994) AUSTRIAN-BRITISH PHILOSOPHER

[The intolerant] may forbid their followers to listen to anything as deceptive as rational argument, and teach them to answer arguments by the use of their fists. We should therefore claim, in the name of tolerance, the right not to tolerate the intolerant.[725]
KARL POPPER (1902-1994) AUSTRIAN-BRITISH PHILOSOPHER

The presence of those seeking the truth is infinitely to be preferred to the presence of those who think they've found it.[726]
TERRY PRATCHETT (1948-2015) ENGLISH WRITER

Man is the measure of all things.[727]
PROTAGORAS (C.481-C.420 BCE) GREEK PHILOSOPHER

A woman is nobody. A wife is everything.[728] [An all-too-com-

mon reaction, even today, to women's rights movements.]
PUBLIC LEDGER AND DAILY TRANSCRIPT (1848) AMERICAN NEWSPAPER

When you look at organized religion of whatever sort - whether it's Christianity in all its variants, or whether it's Islam or some forms of extreme Hinduism - wherever you see organized religion and priesthoods and power, you see cruelty and tyranny and repression. It's almost a universal law.[729]
PHILLIP PULLMAN (1946-) ENGLISH WRITER

For you is your religion, and for me is my religion.[730]
THE QUR'AN

When people use religion as their only reason for whatever laws they want to impose on people or on other things, these are always mostly unjust.[731]
ARON RA (1962-) AMERICAN ACTIVIST

The zealots imposing their religion in public schools never understand that they're even doing anything wrong. They think they're being oppressed or attacked if they're disallowed from indoctrinating everyone else's kids.[732]
ARON RA (1962-) AMERICAN ACTIVIST

It is not God that is worshipped but the group or the authority that claims to speak in his name. Sin becomes disobedience to authority and not violation of integrity.[733]
SARVEPALLI RADHAKRISHNAN (1888-1975) INDIAN PHILOSOPHER

A culture is made - or destroyed - by its articulate voices.[734]
AYN RAND (1905-1982) AMERICAN WRITER

A gun is not an argument.[735]

AYN RAND (1905-1982) AMERICAN WRITER

To each according to his threat advantage does not count as a principle of justice.[736]

JOHN RAWLS (1921-2002) AMERICAN PHILOSOPHER

An intolerant sect has no right to complain when it is denied an equal liberty.[737]

JOHN RAWLS (1921-2002) AMERICAN PHILOSOPHER

Interviewer: *What do you think happens when we die?*
Keanu: *The ones who love us will miss us.*[738]

KEANU REEVES (1964-) CANADIAN ACTOR

A thinking woman sleeps with monsters.[739] [Reflecting on the gender that instigates the most violence.]

ADRIENNE RICH (1929-2012) AMERICAN POET

When you understand why you dismiss all the other possible gods, you will understand why I dismiss your god. I contend we are both atheists. I just believe in one fewer god than you do.[740]

STEPHEN ROBERTS

The feminist agenda is not about equal rights for women. It is about a socialist, anti-family political movement that encourages women to leave their husbands, kill their children, practice witchcraft, destroy capitalism and become lesbians.[741] [There are no words.]

PAT ROBERTSON (1930-) AMERICAN TELEVANGELIST

We hardly find any persons of good sense save those who agree with us.[742]

FRANÇOIS DE LA ROCHEFOUCAULD (1613-1680) FRENCH WRITER

We must question the logic of having an all-knowing, all-powerful God who creates faulty humans, and then blames them for his own mistakes.[743]

GENE RODDENBERRY (1921–1991) AMERICAN WRITER

It wasn't until I was beginning to do Star Trek that the subject of religion arose again. What brought it up was that people were saying that I would have to have a chaplain on board the Enterprise. I replied, "No, we don't."[744]

GENE RODDENBERRY (1921–1991) AMERICAN WRITER

I never cease being dumbfounded by the unbelievable things people believe.[745]

LEO ROSTEN (1908–1997) AMERICAN EDUCATOR

Christianity preaches only servitude and dependence. Its spirit is so favourable to tyranny that it always profits by such a regime. True Christians are made to be slaves, and they know it and do not much mind: this short life counts for too little in their eyes.[746]

JEAN-JACQUES ROUSSEAU (1712–1778) FRENCH PHILOSOPHER

Tolerance should be given to all religions that tolerate others.[747]

JEAN-JACQUES ROUSSEAU (1712–1778) FRENCH PHILOSOPHER

The undertakings which bind us to the social body are obligatory only because they are mutual; and their nature is such that in fulfilling them we cannot work for others without working for ourselves.[748]

JEAN-JACQUES ROUSSEAU (1712–1778) FRENCH PHILOSOPHER

Evolution is a fact, fact, FACT![749]

MICHAEL RUSE (1940–) CANADIAN PHILOSOPHER

The idea of the sacred is quite simply one of the most conservative notions in any culture, because it seeks to turn other ideas - uncertainty, progress, change - into crimes.[750]

SALMAN RUSHDIE (1947-) BRITISH WRITER

The True Believer knows that he is simply right, and you are wrong. He will seek to convert you, even by force, and if he cannot he will, at the very least, despise you for your unbelief.[751]

SALMAN RUSHDIE (1947-) BRITISH WRITER

From the beginning men used God to justify the unjustifiable.[752]

SALMAN RUSHDIE (1947-) BRITISH WRITER

Becoming a purist is commonly indicative of some slight degree of weakness, readiness to be offended, or narrowness of understanding of the ends of things.[753] [Remarking on art, but employing vocabulary equally descriptive of those who believe they, and only they, are in possession of the truth.]

JOHN RUSKIN (1819-1900) ENGLISH POLYMATH

The fact that an opinion has been widely held is no evidence whatever that it is not entirely absurd.[754]

BERTRAND RUSSELL (1872-1970) BRITISH PHILOSOPHER

Persecution is used in theology, not in arithmetic, because in arithmetic there is knowledge, but in theology there is only opinion.[755]

BERTRAND RUSSELL (1872-1970) BRITISH PHILOSOPHER

The Christian religion has been and still is the principal enemy of moral progress in the world.[756]

BERTRAND RUSSELL (1872-1970) BRITISH PHILOSOPHER

Ethics is in origin the art of recommending to others the sacrifices required for co-operation with oneself.[757]

BERTRAND RUSSELL (1872–1970) BRITISH PHILOSOPHER

The authority of science, which is recognized by most philosophers of the modern epoch, is a very different thing from the authority of the Church, since it is intellectual, not governmental. No penalties fall upon those who reject it; no mystical arguments influence those who accept it. It prevails solely by its intrinsic appeal to reason. It is, moreover, a piecemeal and partial authority; it does not, like the bodies of religious dogma, lay down a complete system, covering human morality, human hopes, and the past and future history of the universe. It pronounces only on whatever, at the time, appears to have been scientifically ascertained, which is a small island in an ocean of nescience. There is yet another difference from ecclesiastical authority, which declares its pronouncements to be absolutely certain and eternally unalterable: the pronouncements of science are made tentatively, on a basis of probability, and are regarded as liable to modification. This produces a temper of mind very different from that of the medieval [or present-day] dogmatist.[758]

BERTRAND RUSSELL (1872–1970) BRITISH PHILOSOPHER

Even if [a] dogmatically held opinion should happen to be true, the mere fact that it is held dogmatically will cause it to be harmful.[759]

BERTRAND RUSSELL (1872–1970) BRITISH PHILOSOPHER

Religions are tough. Either they make no contentions which are subject to disproof or they quickly redesign doctrine after disproof. The fact that religions can be so shamelessly dishonest, so contemptuous of the intelligence of their adherents, and still flourish does not speak very well of the tough-mindedness of the believers. But it does indicate, if a demonstration were needed, that near the core of the religious experience is something remarkably resistant to rational inquiry.[760]

CARL SAGAN (1934-1996) AMERICAN PHYSICIST

Authorities must prove their contentions like everybody else.[761]

CARL SAGAN (1934-1996) AMERICAN PHYSICIST

Our posturings, our imagined self-importance, the delusion that we have some privileged position in the universe - all are challenged by the immateriality of our unassuming planet, a mote of dust suspended in a sunbeam, a pale point of light in the great cosmic dark.[762]

CARL SAGAN (1934-1996) AMERICAN PHYSICIST

In science it often happens that scientists say, "You know, that's a really good argument; my position is mistaken," and then they would actually change their minds and you never hear that old view from them again. They really do it. It doesn't happen as often as it should, because scientists are human and change is sometimes painful. But it happens every day. I cannot recall the last time something like that happened in politics or religion.[763]

CARL SAGAN (1934-1996) AMERICAN PHYSICIST

There are no forbidden questions in science.[764]

CARL SAGAN (1934-1996) AMERICAN PHYSICIST

Intellectual freedom is essential to human society - freedom to obtain and distribute information, freedom for open-minded and unfearing debate, and freedom from pressure by officialdom and prejudices.[765]

ANDREI SAKHAROV (1921-1989) RUSSIAN PHYSICIST

Religion has never befriended the cause of humaneness. Its monstrous doctrine of eternal punishment and the torture of the damned underlies much of the barbarity with which man has treated man.[766]

HENRY STEPHENS SALT (1851–1939) ENGLISH WRITER

I would prefer to believe that God does not exist rather than believe He is indifferent.[767]

GEORGE SAND (1804–1876) FRENCH WRITER

The church has ever opposed the progress of woman on the ground that her freedom would lead to immorality. We ask the church to have more confidence in women.[768]

MARGARET SANGER (1879–1966) AMERICAN ACTIVIST

No gods, no masters.[769]

MARGARET SANGER (1879–1966) AMERICAN ACTIVIST

Each religion ... necessarily contradicts every other religion, and probably contradicts itself.[770]

GEORGE SANTAYANA (1863–1952) SPANISH PHILOSOPHER

The sciences are necessarily allies, but religions, like languages, are necessarily rivals.[771]

GEORGE SANTAYANA (1863–1952) SPANISH PHILOSOPHER

Christianity persecuted, tortured, and burned. Like a hound it tracked the very scent of heresy.[772]

GEORGE SANTAYANA (1863–1952) SPANISH PHILOSOPHER

The most malicious kind of hatred is that which is built upon a theological foundation.[773]

GEORGE SARTON (1884–1956) AMERICAN HISTORIAN

The anti-Semite has chosen hate because hate is a faith; at the outset he has chosen to devaluate words and reasons.[774]

JEAN-PAUL SARTRE (1905-1980) FRENCH PHILOSOPHER

The character of every act depends upon the circumstances in which it is done.[775] [Commenting on the existence of few, if any, legal or moral absolutes.]

SCHENCK V UNITED STATES (1919) UNITED STATES SUPREME COURT

Those who injure the souls of believers with scorn and mockery must be put in their place and, if necessary, also be punished.[776] [A reminder of religion's intrinsic stance on freedom of speech.]

LUDWIG SCHICK (1949-) GERMAN THEOLOGIAN

There are people whose sole aspiration is to say no. They wield their acerbity like a sharp pair of scissors, purging the extremities of genius; their enlightenment is that of the candle-snuffer, intent on extinguishing the flame of enthusiasm.[777]

FRIEDRICH SCHLEGEL (1772-1829) GERMAN POET

As a historian, I confess to a certain amusement when I hear the Judeo-Christian tradition praised as the source of our concern for human rights. In fact, the great religious ages were notable for their indifference to human rights in the contemporary sense. They were notorious not only for acquiescence in poverty, inequality, exploitation and oppression but for enthusiastic justifications of slavery, persecution, abandonment of small children, torture, genocide.[778]

ARTHUR SCHLESINGER, JR. (1917-2007) AMERICAN HISTORIAN

Human rights is not a religious idea. It is a secular idea, the product of the last four centuries of Western history.[779]

ARTHUR SCHLESINGER, JR. (1917-2007) AMERICAN HISTORIAN

The absurdities of dogma are marks of allegory and myth.[780]

ARTHUR SCHOPENHAUER (1788-1860) GERMAN PHILOSOPHER

All religions and philosophies promise a reward in eternity only to the virtues of the will, or heart, not to those of the intellect or the head.[781]

ARTHUR SCHOPENHAUER (1788-1860) GERMAN PHILOSOPHER

Let me give you a definition of ethics: It is good to maintain and further life - it is bad to damage and destroy life.[782]

ALBERT SCHWEITZER (1875-1965) ALSATIAN THEOLOGIAN

A great many persons' identities as Indians, as subcontinentals, as Asians, or as members of the human race, seemed to give way - quite suddenly - to sectarian identification with Hindu, Muslim or Sikh communities. The carnage that followed had much to do with elementary herd behavior by which people were made to "discover" their newly detected belligerent identities, without subjecting the process to critical examination.[783] [An early condemnation of identity politics.]

AMARTYA SEN (1933-) INDIAN ECONOMIST

He who gains by a crime, committed it.[784]

SENECA THE YOUNGER (C.4 BCE-65 CE) ROMAN PHILOSOPHER

I do not in all points agree, nor disagree, with either one party or the other. All seem to me to have some truth and some error, and every one perceives the other's error [but] no one sees his own.[785]

MICHAEL SERVETUS (1511-1553) SPANISH THEOLOGIAN

The fault, dear Brutus, is not in our stars but in ourselves.[786]

WILLIAM SHAKESPEARE (1564-1616) ENGLISH PLAYWRIGHT

Imagine your son becomes your guardian. No matter my capabilities as a woman, I am still enslaved to somebody else. Freedom for me is to live with dignity, and if my dignity and freedom is controlled by a man, I will never be free.[787]

MANAL AL-SHARIF (1979-) SAUDI ACTIVIST

All great truths begin as blasphemies.[788]

GEORGE BERNARD SHAW (1856-1950) IRISH PLAYWRIGHT

It is not disbelief that is dangerous to our society; it is belief.[789]

GEORGE BERNARD SHAW (1856-1950) IRISH PLAYWRIGHT

For Catholics before Vatican II, the land of the free was preeminently the land of Sister Says - except, of course, for Sister, for whom it was the land of Father Says.[790]

WILFRID SHEED (1930-2011) ENGLISH-AMERICAN WRITER

I will admit that one prediction of Jesus Christ has been indisputably fulfilled:[791] *"Think not that I am come to send peace on earth: I came not to send peace, but a sword."*[792]

PERCY BYSSHE SHELLEY (1792-1822) ENGLISH POET

You should not use violence; ... the courtyard will hear of it.[793] [Advice, oft-repeated but oft-ignored, from one of the earliest known texts.]

THE INSTRUCTIONS OF SHURUPPAG (C.2600 BCE) SUMERIAN TEXT

There are a score of great religions in the world, each with scores or hundreds of sects, each with its priestly orders, its complicated creed and ritual, its heavens and hells. Each has its thousands or millions or hundreds of millions of "true believers"; each damns all the others with more or less heartiness - and each is a mighty fortress of graft.[794]

UPTON SINCLAIR (1878–1968) AMERICAN WRITER

Human beings are social animals. We were social before we were human.[795]

PETER SINGER (1946–) AUSTRALIAN PHILOSOPHER

I felt I had to choose between Darwin and the Bible. ... There's no contest, you have to accept Darwin's view. The god-creation view is nonsense.[796] [On the relation between deities and the theory of natural selection as explanatory paradigms.]

JOHN MAYNARD SMITH (1920–2004) BRITISH GENETICIST

Belief in Some One's right to punish you is the fate of all children in Judaic-Christian culture. But nowhere else, perhaps, have the rich seedbeds of Western homes found such a growing climate for guilt as is produced in the [American] South by the combination of a warm, moist evangelism and racial segregation.[797]

LILIAN SMITH (1897–1966) AMERICAN WRITER

The common assumption, hardly disputed even now, that the moral influence of the Bible has been wholly good, and that all that is needed to improve our society is to "spread the gospel," is not borne out by a candid study of history.[798]

PRESERVED SMITH (1880–1941) AMERICAN HISTORIAN

We hear with surprise of the savage who, falling down a precipice, ascribes the failure of his foothold to a malicious demon. But daily, without surprise, we hear the Christian priest who says prayers over a sick man in the expectation that the course of his disease will be stayed. The savage and the priest differ only in respect of the agent from which they expect supernatural aid.[799]

HERBERT SPENCER (1820-1903) ENGLISH PHILOSOPHER

Men who are governed by reason desire nothing for themselves which they do not desire for other men.[800]

BENEDICT DE SPINOZA (1632-1677) DUTCH PHILOSOPHER

Burning stakes do not lighten the darkness.[801]

LEC STANISLAW (1909-1966) POLISH POET

It is the high priests that make demands - not the gods they serve.[802]

LEC STANISLAW (1909-1966) POLISH POET

The lessons of self-sacrifice and obedience taught woman by the Christian church have been fatal, not only to her vital interests, but through her, to those of the race.[803]

ELIZABETH CADY STANTON (1815-1902) AMERICAN ACTIVIST

The Bible teaches that woman brought sin and death into the world, that she precipitated the fall of the race, that she was arraigned before the judgment seat of Heaven, tried, condemned and sentenced. Marriage for her was to be a condition of bondage, maternity a period of suffering and anguish, and in silence and subjection, she was to play the role of a dependent on man's bounty for all her material wants, and for all the information she might desire on the vital questions of the hour, she was commanded to ask her husband at home. Here is the Bible position of woman briefly summed up.[804]

ELIZABETH CADY STANTON (1815-1902) AMERICAN ACTIVIST

I have lived long enough to see that differences beget hatred.[805]

STENDHAL (1783-1842) FRENCH WRITER

What chiefly concerns and alarms many of us are the problems arising from religious fanaticism. As long as large numbers of militant enthusiasts are persuaded that they alone have access to the truth, and that the rest of us are infidels, we remain under threat.[806]

ANTHONY STORR (1920–2001) ENGLISH PSYCHIATRIST

The number of those men who know how to use wholly irresponsible power humanely and generously is small.[807]

HARRIET BEECHER STOWE (1811–1896) AMERICAN ABOLITIONIST

Heresy is the lifeblood of religions. It is faith that begets heretics. There are no heresies in a dead religion.[808]

ANDRÉ SUARÈS (1868–1948) FRENCH POET

It is useless to attempt to reason a man out of a thing he was never reasoned into.[809]

JONATHAN SWIFT (1667–1745) IRISH WRITER

Violent zeal for truth hath a hundred to one odds to be either petulant, ambition, or pride.[810]

JONATHAN SWIFT (1667–1745) IRISH WRITER

The intention to injure can always find a reason.[811]

PUBLILIUS SYRUS (84–43 BCE) LATIN WRITER

When [totalitarian] leaders jail a writer, or a priest, or a doctor, or a worker, for the crime of speaking freely, it is not only for humanitarian reasons that we should be concerned. For these acts reveal a regime that is afraid of truth and liberty; it dare not allow its people to enjoy the freedoms we take for granted, and a nation that denies those freedoms to its own people will have few scruples in denying them to others.[812]

MARGARET THATCHER (1925-2013) BRITISH PRIME MINISTER

In politics, if you want anything said ask a man, if you want anything done ask a woman.[813]

MARGARET THATCHER (1925-2013) BRITISH PRIME MINISTER

If we go back to the origin of things, we shall always find that ignorance and fear created the gods; that imagination, enthusiasm and deception adorned or disfigured them; that weakness adores them; that credulity nourishes them; that habit preserves them; that tyranny sustains them - all in order to profit from the blindness of men.[814]

PAUL-HENRI THIRY (1723-1789) FRENCH PHILOSOPHER

It is believed that those who deny the existence of God can be silenced by telling them that all men, in all ages, in all countries, have recognized the empire of some divine power. But can the belief of all men transform an error into a truth?[815]

PAUL-HENRI THIRY (1723-1789) FRENCH PHILOSOPHER

Little pains do the vulgar take in the investigation of truth, accepting readily the first story that comes to hand.[816]

THUCYDIDES (C.460-C.400 BCE) GREEK HISTORIAN

Before God and on the Holy Gospel I swear and promise, as becomes a bishop, loyalty to the German Reich.[817] [Oath required by Catholic bishops in exchange for nonintervention by the state in ecclesiastical affairs. The Church came to regret the pact.]

TREATY BETWEEN THE HOLY SEE AND THE GERMAN REICH (1933)

The reasonableness of something is rarely a motive in human affairs.[818]

BARBARA TUCHMAN (1912-1989) AMERICAN HISTORIAN

For most people reform meant relief from ecclesiastical extortions.[819] [Commenting on the behavior of the Church in the Middle Ages.]

BARBARA TUCHMAN (1912-1989) AMERICAN HISTORIAN

Many ... people have the reasoning faculty, but no one uses it in religious matters.[820]

MARK TWAIN (1835-1910) AMERICAN WRITER

Everyone has the right to freedom of thought, conscience and religion; this right includes freedom to change his religion or belief, and freedom, either alone or in community with others and in public or private, to manifest his religion or belief in teaching, practice, worship and observance.[821]

UNIVERSAL DECLARATION OF HUMAN RIGHTS (1948) ARTICLE 18

Everyone has the right to freedom of opinion and expression; this right includes freedom to hold opinions without interference and to seek, receive and impart information and ideas through any media and regardless of frontiers.[822]

UNIVERSAL DECLARATION OF HUMAN RIGHTS (1948) ARTICLE 19

Every sect is a rallying point for doubt and error.[823]

VOLTAIRE (1694-1778) FRENCH PHILOSOPHER

If God did not exist, it would be necessary to invent him.[824]

VOLTAIRE (1694-1778) FRENCH PHILOSOPHER

I must tell you what happened to me one day. I heard a mole who was reasoning with a beetle. "Here is a fine structure," said the mole. "A very powerful mole must have built it." "You must be joking," responded the beetle. "It could only have been a beetle of great genius that constructed it."[825] [On our tendency to fabricate gods with human visages.]

VOLTAIRE (1694-1778) FRENCH PHILOSOPHER

I am a humanist, which means, in part, that I have tried to behave decently without any expectation of rewards or punishments after I'm dead.[826]

KURT VONNEGUT (1922-2007) AMERICAN WRITER

Imagine a world in which every single person on the planet has free access to the sum of all human knowledge.[827]

JIMMY WALES (1966-) WIKIPEDIA CO-FOUNDER

I learned the lesson that great men cultivate love, and that only little men cherish a spirit of hatred.[828]

BOOKER T. WASHINGTON (1856-1915) AMERICAN EDUCATOR

I see nothing in this proceeding, but good policy, combined with mercy.[829] [Endorsing the ancient command of his deity to slay the Midianites, enslave their maidens, and burn their cities.[830]]

RICHARD WATSON (1737-1816) ENGLISH THEOLOGIAN

I have noticed again and again since I have been in the Church that lay interest in ecclesiastical matters is often a prelude to insanity.[831]

EVELYN WAUGH (1903-1966) ENGLISH WRITER

The human soul has need of truth and of freedom of expression.[832]

SIMONE WEIL (1909-1943) FRENCH PHILOSOPHER

I think the world needs to wake up from its long nightmare of religious belief; and anything that we scientists can do to weaken the hold of religion should be done, and may in fact be our greatest contribution to civilization.[833]

STEVEN WEINBERG (1933-) AMERICAN PHYSICIST

What hurts the victim most is not the cruelty of the oppressor but the silence of the bystander.[834]

ELIE WIESEL (1928-2016) AMERICAN WRITER

All collective judgments are wrong. Only racists make them. And racism is stupid, just as it is ugly. Its aim is to destroy, to pervert, to distort innocence in human beings and their quest for human equality. Racism is misleading. There are good people and bad people in every community. No human race is superior; no religious faith is inferior.[835]

ELIE WIESEL (1928-2016) AMERICAN WRITER

Increasing knowledge lessens the sphere of the supernatural.[836]

EDWARD WESTERMARCK (1862-1939) FINNISH PHILOSOPHER

Religion is the last refuge of human savagery.[837]

ALFRED NORTH WHITEHEAD (1861-1947) ENGLISH PHILOSOPHER

Lying, the telling of beautiful untrue things, is the proper aim of art.[838]

OSCAR WILDE (1854-1900) IRISH PLAYWRIGHT

Morality is simply the attitude we adopt towards people whom we personally dislike.[839]

OSCAR WILDE (1854-1900) IRISH PLAYWRIGHT

When I think of all the harm the Bible has done, I despair of

ever writing anything equal to it.[840]
OSCAR WILDE (1854-1900) IRISH PLAYWRIGHT

All fanaticism is a strategy to prevent doubt from becoming conscious.[841]
A. H. WILLIAMS (1862-1939) FINNISH PHILOSOPHER

I can't get rid of the idea that somewhere in the universe there must be a creature superior to man.[842]
VOICED BY THE CHARACTER TAYLOR (1968) PLANET OF THE APES

The real problem of humanity is the following: we have Paleolithic emotions; medieval institutions; and god-like technology.[843]
E. O. WILSON (1929-) AMERICAN BIOLOGIST

It is justice, not charity, that is wanting in the world.[844]
MARY WOLLSTONECRAFT (1759-1797) ENGLISH WRITER

The white folks like for us to be religious, then they can do what they want with us.[845]
RICHARD WRIGHT (1908-1960) AMERICAN WRITER

[In the United States] it is shameful that there are so few women in science. In China there are many, many women in physics. There is a misconception in America that women scientists are all dowdy spinsters. This is the fault of men.[846]
CHIEN-SHIUNG WU (1912-1997) CHINESE-AMERICAN PHYSICIST

When women are well-educated, men benefit.[847]
CHIEN-SHIUNG WU (1912-1997) CHINESE-AMERICAN PHYSICIST

Good thoughts, good words, good deeds.[848] [Admirable principles, within any philosophy or belief system.]

ZEND-AVESTA (CIRCA 200-700 CE) ZOROASTRIAN SACRED TEXT

I am supposing, or perhaps only hoping, that our future may be found in the past's fugitive moments of compassion rather than in its solid centuries of warfare.[849]

HOWARD ZINN (1924-2010) AMERICAN HISTORIAN

It is the great challenge of our time: How to achieve justice, with struggle, but without war.[850] [It is the great challenge of all times.]

HOWARD ZINN (1924-2010) AMERICAN HISTORIAN

MISCELLANY

Humanists with an orthographic bent suggest that the word *humanism* should not be capitalized, as the philosophy is not an ideology or a religion. I bow to this position, presenting the term without capitals. Describing Enlightenment humanism as a *philosophy*, however, might raise some hairs. Many humanists avoid the term, due in part to its metaphysical and speculative connotations. Instead, they label the philosophy a worldview, a life stance, or simply a way of thinking. Yet it all amounts to much the same. Philosophy is a fine term, and its etymology favors its use. It derives from Greek, with *philos* (*φίλος*) denoting *friend* and *sophia* (*γνώση*) meaning *knowledge*. Together, they form *friend of knowledge*, a concept that stands in perfect alignment with Enlightenment humanism.

ON DEMOCRACY

The best test of truth is the power of thought to get itself accepted in the competition of the market. ... That, at any rate, is the theory of our Constitution. It is an experiment, as all life is an experiment.[851]

ABRAMS V UNITED STATES (1919) UNITED STATES SUPREME COURT

Power tends to corrupt, and absolute power corrupts abso-

lutely.[852]

JOHN DALBERG-ACTON (1834-1902) ENGLISH HISTORIAN

The most certain test by which we judge whether a country is really free is the amount of security enjoyed by minorities.[853]

JOHN DALBERG-ACTON (1834-1902) ENGLISH HISTORIAN

Unequal electorates afford no security to majorities. Equal electorates give none to minorities.[854]

JOHN DALBERG-ACTON (1834-1902) ENGLISH HISTORIAN

The liberty of the press is essential to the security of freedom in a state.[855]

JOHN ADAMS (1735-1826), SAMUEL ADAMS (1722-1803), AND JAMES BOWDOIN (1726-1790) AMERICAN STATESMEN

Children should be educated and instructed in the principles of freedom.[856]

JOHN ADAMS (1735-1826) AMERICAN PRESIDENT

It is in truth not for glory, nor riches, nor honours, that we are fighting, but for freedom - for that alone, which no honest man gives up but with life itself.[857]

THE DECLARATION OF ARBROATH (1320) SCOTTISH PROCLAMATION

Theocracy has always been the synonym for a bleak and narrow, if not a fierce blood-stained, tyranny.[858]

WILLIAM ARCHER (1856-1924) SCOTTISH WRITER

Before mass leaders seize the power to fit reality to their lies, their propaganda is marked by its extreme contempt for facts as such, for in their opinion fact depends entirely on the power of [the] man who can fabricate it.[859]

HANNAH ARENDT (1906-1975) GERMAN-AMERICAN POLITICAL THEORIST

The totalitarian mass leaders base their propaganda on the correct psychological assumption that, under such conditions, one could make people believe the most fantastic statements one day, and trust that if the next day they were given irrefutable proof of their falsehood, they would take refuge in cynicism.[860]

HANNAH ARENDT (1906-1975) GERMAN-AMERICAN POLITICAL THEORIST

The ideal subject of totalitarian rule is not the convinced adherent of the far-right or the far-left, but people for whom the distinction between fact and fiction, and the distinction between true and false, no longer exists.[861]

HANNAH ARENDT (1906-1975) GERMAN-AMERICAN POLITICAL THEORIST

For if liberty and equality, as is thought by some, are chiefly to be found in democracy, they will be best attained when all persons alike share in the government to the utmost.[862]

ARISTOTLE (384-322 BCE) GREEK PHILOSOPHER

It is manifest that the best political community is formed by citizens of the middle class, and that those states are likely to be well-administered in which the middle class is large and strong.[863] [Opining not only on the ingredients necessary to form and sustain a democracy but, equally, how to make its citizenry "most ready to follow the rational principle".[864]

ARISTOTLE (384-322 BCE) GREEK PHILOSOPHER

Democracy is the false notion that my ignorance is just as good as your knowledge.[865]

ISAAC ASIMOV (1920-1992) AMERICAN WRITER

As soon as a thinker starts to reveal his ideas, you will find hun-

dreds of fatwas that accuse him of being an infidel.[866]
RAIF BADAWI (1984-) SAUDI WRITER

The idea of God implies the abdication of human reason and justice; it is the most decisive negation of human liberty, and necessarily ends in the enslavement of mankind, both in theory and practice.[867]
MIKHAIL BAKUNIN (1814-1876) RUSSIAN ACTIVIST

The concepts contained in words like "freedom," "justice," "democracy" are not common concepts; on the contrary, they are rare. People are not born knowing what these are. It takes enormous and, above all, individual effort to arrive at the respect for other people that these words imply.[868]
JAMES BALDWIN (1924-1987) AMERICAN WRITER

If one really wishes to know how justice is administered in a country, one does not question the policemen, the lawyers, the judges, or the protected members of the middle class. One goes to the unprotected.[869]
JAMES BALDWIN (1924-1987) AMERICAN WRITER

Political liberty, the peace of a nation, and science itself are gifts for which fate has laid a tax of blood.[870]
HONORÉ DE BALZAC (1799-1850) FRENCH WRITER

No child should have to die for going to school.[871] [Commenting on the wounding of a Pakistani schoolgirl who was shot by religious extremists for wanting an education.]
BAN KI-MOON (1944-) SECRETARY GENERAL OF THE UNITED NATIONS

Where the people fear the government you have tyranny. Where the government fears the people you have liberty.[872]

JOHN BASIL BARNHILL (1864-1929) AMERICAN WRITER

If [the devout] would duly consider what is to be feared from a religious war, they would be moderate.[873]

PIERRE BAYLE (1647-1706) FRENCH PHILOSOPHER

Let liberty be attended with knowledge.[874]

CESARE BECCARIA (1738-1794) ITALIAN JURIST

The primal principle of democracy is the worth and dignity of the individual.[875]

EDWARD BELLAMY (1850-1898) AMERICAN WRITER

Few things have done more harm than the belief on the part of individuals or groups (or tribes or states or nations or churches) that he or she or they are in the sole possession of the truth ... and that those who differ from them are not merely mistaken, but wicked or mad, and need restraining or suppressing.[876]

ISAIAH BERLIN (1909-1997) RUSSIAN-BRITISH PHILOSOPHER

Democracies cannot dispense with hypocrisy any more than dictatorship can with cynicism.[877]

GEORGES BERNANOS (1888-1948) FRENCH WRITER

Anyone who is indifferently open to truth or falsehood is ripe for any kind of tyranny.[878]

GEORGES BERNANOS (1888-1948) FRENCH WRITER

I represent everything [the Taleban and al-Qaeda] fear the most - moderation, democracy, equality for women, information, and technology.[879]

BENAZIR BHUTTO (1953-2007) PAKISTANI PRIME MINISTER

Politics: A strife of interests masquerading as a contest of principles.[880]

AMBROSE BIERCE (1842-C.1914) AMERICAN SATIRIST

Nothing ... is to be more avoided, in a free constitution, than uniting the provinces of a judge and a minister of state.[881]

WILLIAM BLACKSTONE (1723-1780) ENGLISH JURIST

The continuation of authority in the same person has frequently proved the undoing of democratic governments. The people become accustomed to obeying him, and he becomes accustomed to commanding, hence the origin of usurpation and tyranny.[882]

SIMÓN BOLIVAR (1783-1830) VENEZUELAN INDEPENDENCE LEADER

Every liberty won by mankind has been won in spite of the hostility of the Church in power, not through its aid.[883]

HYPATIA BRADLAUGH BONNER (1858-1935) BRITISH ACTIVIST

Royal authority is sacred[884] *and absolute.*[885] [Affirming the historical bond between religion and despotic power.]

JACQUES-BÉNIGNE BOSSUET (1627-1704) FRENCH THEOLOGIAN

No government can long endure unless its citizens are willing to make some sacrifice for its existence.[886]

WILLIAM JENNINGS BRYAN (1860-1925) AMERICAN STATESMAN

Conservatives pride themselves on resisting change, which is as it should be. But intelligent deference to tradition and stability can evolve into intellectual sloth and moral fanaticism, as when conservatives simply decline to look up from dogma because the effort to raise their heads and reconsider is too great.[887] [Buckley, on conservatives.]

WILLIAM F. BUCKLEY, JR. (1925-2008) AMERICAN WRITER

Though liberals do a great deal of talking about hearing other points of view, it sometimes shocks them to learn that there ARE other points of view.[888] [Buckley, on liberals.]

WILLIAM F. BUCKLEY, JR. (1925-2008) AMERICAN WRITER

Socialism takes away a man's cow but leaves him the milk; fascism leaves him the cow but takes away the milk; communism takes away both cow and milk.[889]

IVOR BULMER-THOMAS (1905-1993) BRITISH JOURNALIST

It is a general popular error to imagine the loudest complainers for the public to be the most anxious for its welfare.[890]

EDMUND BURKE (1729-1797) IRISH PHILOSOPHER

Education is the cheap defence of nations.[891]

EDMUND BURKE (1729-1797) IRISH PHILOSOPHER

Nothing should be left undone to impress upon the young that freedom of thought is an axiom of human progress.[892]

JOHN BAGNELL BURY (1861-1927) IRISH HISTORIAN

It is harder to preserve than to obtain liberty.[893]

JOHN C. CALHOUN (1782-1850) AMERICAN STATESMAN

There always comes a time in history when the person who dares to say that 2 + 2 = 4 is punished by death.[894]

ALBERT CAMUS (1913-1960) FRENCH WRITER

Democracy is the power of equal votes for unequal minds.[895] [A royalist's view of democracy.]

CHARLES I (1600-1649) KING OF ENGLAND, IRELAND AND SCOTLAND

Every man objects to any despot but himself.[896]

MARY BOYKIN CHESNUT (1823-1886) AMERICAN WRITER

The smart way to keep people passive and obedient is to strictly limit the spectrum of acceptable opinion, but allow very lively debate within that spectrum - even encourage the more critical and dissident views. That gives people the sense that there's free thinking going on, while all the time the presuppositions of the system are being reinforced by the limits put on the range of the debate.[897]

NOAM CHOMSKY (1928-) AMERICAN PHILOSOPHER

Democracy is not based on violence or terrorism, but on reason, on fair play, on freedom, on respecting the rights of other people.[898]

WINSTON CHURCHILL (1600-1649) BRITISH PRIME MINISTER

It has been said that democracy is the worst form of government, except all those other forms that have been tried.[899]

WINSTON CHURCHILL (1600-1649) BRITISH PRIME MINISTER

If we are so disposed that each, to gain some personal profit, will defraud or injure his neighbor, then those bonds of human society ... must of necessity be broken.[900]

CICERO (106-43 BCE) ROMAN STATESMAN

My son is 22 years old. If he had not become a Communist at 22, I would have disowned him. If he is still a Communist at 30, I will do it then.[901]

GEORGES CLEMENCEAU (1841-1929) FRENCH PRIME MINISTER

It is in one word, the only word that ever brought equity anywhere - liberty.[902]

VOLTAIRINE DE CLEYRE (1866-1912) AMERICAN ACTIVIST

There is no just remedy to [superstition] but free-thinking.[903]

ANTHONY COLLINS (1676-1729) ENGLISH PHILOSOPHER

Child, Satan desires to have you, and so he is putting such questions into your head; answer him as did the Saviour, "Get thee behind me, Satan!" and remember it is very wicked to reason on the way of God; you have the Holy Bible, read that, and accept it.[904] [Colman, recalling the responses she received whenever she questioned religious doctrine.]

LUCY COLMAN (1817-1906) AMERICAN ABOLITIONIST

Christianity demands entire subordination to its edicts.[905]

LUCY COLMAN (1817-1906) AMERICAN ABOLITIONIST

Habit can familiarize men with the violation of natural rights. For example, have not all violated the principle of equal rights, by quietly depriving half of the human race of the right to contribute to the formation of laws, by excluding women? For this exclusion not to be an act of tyranny, it would be necessary either to prove that the natural rights of women are not absolutely the same as those of men, or to show that they are not capable of exercising them.[906]

MARQUIS DE CONDORCET (1743-1794) FRENCH PHILOSOPHER

Dictators seek to control men's thoughts as well as their bodies and so they attempt to dictate science, education and religion. But dictated education is usually propaganda, dictated history is often mythology, dictated science is pseudo-science.[907]

EDWIN GRANT CONKLIN (1863-1952) AMERICAN BIOLOGIST

When you put on your Jesus glasses, you can't see the truth.[908]

JAMES CORBETT (?) AMERICAN EDUCATOR

I dissolve this Parliament.[909] [When religious authority has power over popular government.]

OLIVER CROMWELL (1599-1658) ENGLISH PURITAN

You can only protect your liberties in this world by protecting the other man's freedom. You can only be free if I am free.[910]

CLARENCE DARROW (1857-1938) AMERICAN LAWYER

The origin of what we call civilization is not due to religion, but to skepticism.[911]

CLARENCE DARROW (1857-1938) AMERICAN LAWYER

In every age it has been the tyrant in the cloak of patriotism, or religion, or both, to deceive and overawe the people.[912]

EUGENE DEBS (1855-1926) AMERICAN POLITICIAN

Governments are instituted among men, deriving their just powers from the consent of the governed.

AMERICAN DECLARATION OF INDEPENDENCE (1776)

Intellectually, religious emotions are not creative but conservative. They attach themselves to the current view of the world and consecrate it.[913]

JOHN DEWEY (1859-1952) AMERICAN PHILOSOPHER

"Believe, or you will be damned." This is the strongest argument in theology.[914]

DENIS DIDEROT (1713-1784) FRENCH PHILOSOPHER

From fanaticism to barbarism is but one step.[915]

DENIS DIDEROT (1713-1784) FRENCH PHILOSOPHER

The world is wearied of statesmen, whom democracy has degraded into politicians.[916]

BENJAMIN DISRAELI (1804-1881) BRITISH PRIME MINISTER

Those who profess to favor freedom, and yet depreciate agitation, are men who want crops without plowing up the ground.[917]

FREDERICK DOUGLASS (1818-1895) AMERICAN ABOLITIONIST

All deductions having been made, democracy has done less harm, and more good, than any other form of government.[918]

WILL DURANT (1885-1981) AMERICAN HISTORIAN

Unthinking respect for authority is the greatest enemy of truth.[919]

ALBERT EINSTEIN (1879-1955) GERMAN PHYSICIST

A sect or a party is an elegant incognito devised to save man from the vexation of thinking.[920]

RALPH WALDO EMERSON (1803-1882) AMERICAN WRITER

The educated only are free.[921]

EPICTETUS (C.55-135 CE) GREEK PHILOSOPHER

I am a lover of liberty. I will not and cannot serve a party.[922]

ERASMUS (1466-1536) DUTCH THEOLOGIAN

The "establishment of religion" clause of the First Amendment means at least this: neither a state nor the Federal Government can set up a church. Neither can pass laws which aid one religion, aid all religions, or prefer one religion over another. Neither can force nor influence a person to go to or to remain away from church against his will or force him to profess a belief or disbelief in any religion.[923]

EVERSON V BOARD OF EDUCATION (1947) UNITED STATES SUPREME COURT

It [the First Amendment of the Constitution] was intended not only to keep the states' hands out of religion, but to keep religion's hands off the state, and, above all, to keep bitter religious controversy out of public life by denying to every denomination any advantage from getting control of public policy or the public purse.[924]

EVERSON V BOARD OF EDUCATION (1947) UNITED STATES SUPREME COURT

There is an interesting resemblance in the speeches of dictators, no matter what country they may hail from or what language they may speak.[925]

EDNA FERBER (1885–1968) AMERICAN WRITER

Two cheers for democracy: one because it admits variety and two because it permits criticism. Two cheers are quite enough: there is no occasion to give three.[926]

E. M. FORSTER (1879–1970) ENGLISH WRITER

An inquirer, at the close of the Constitutional Convention of 1787: *Well, what have we got, a republic or a monarchy?*
Franklin: *A republic, if you can keep it.*[927]

BENJAMIN FRANKLIN (1706–1790) AMERICAN STATESMAN

If an exchange between two parties is voluntary, it will not take place unless both believe they will benefit from it. Most economic fallacies derive from the neglect of this simple insight, from the tendency to assume that there is a fixed pie, that one party can gain only at the expense of another.[928]

MILTON FRIEDMAN (1912–2006) AMERICAN ECONOMIST

A society that puts equality ahead of freedom will end up with neither. One that puts freedom first will end up with greater of the both.[929]

MILTON FRIEDMAN (1912–2006) AMERICAN ECONOMIST

When Fascism came into power, most people were unprepared, both theoretically and practically. They were unable to believe that man could exhibit such propensities for evil, such lust for power, such disregard for the rights of the weak, or such yearning for submission.[930]

ERICH FROMM (1900–1980) GERMAN SOCIOLOGIST

Power confuses itself with virtue and tends also to take itself for omnipotence.[931]

J. WILLIAM FULBRIGHT (1905–1995) AMERICAN STATESMAN

Both totalitarian regimes, identical in their aim for absolute power over dehumanized beings, presented themselves as protection from the dangers presented by the other.[932] [Referring to Nazi fascism and Soviet communism.]

FRANÇOIS FURET (1927–1997) FRENCH HISTORIAN

The church and civilization are antipodal; one means authority, the other freedom; one means conservatism, the other progress; one means the rights of God as interpreted by the priesthood, the other the rights of humanity as interpreted by humanity. Civilization advances by free-thought, free speech, free men.[933]

MATILDA JOSLYN GAGE (1826–1898) AMERICAN SUFFRAGIST

Any action that is dictated by fear or by coercion of any kind ceases to be moral.[934]

MOHANDAS GANDHI (1869–1948) INDIAN ACTIVIST

In [an] age of the rule of brute force, it is almost impossible for anyone to believe that any one else could possibly reject the law of ... brute force.[935]

MOHANDAS GANDHI (1869-1948) INDIAN ACTIVIST

Bad science contributes to the steady dumbing down of our nation. Crude beliefs get transmitted to political leaders and the result is considerable damage to society. We see this happening now in the rapid rise of the religious right and how it has taken over large segments of the Republican Party.[936]

MARTIN GARDNER (1914-2010) AMERICAN WRITER

From differences in faith or church, we pledge to shed no blood, to confiscate no property, to threaten no imprisonment or exile, and to assist no authority in any such acts.[937] [Oft-cited as the first declaration of religious tolerance in Europe.]

GENERAL CONFEDERATION OF WARSAW (1573)

The influence of the clergy, in an age of superstition, might be usefully employed to assert the rights of mankind; but so intimate is the connection between the throne and the alter, that the banner of the church has very seldom been seen on the side of the people.[938]

EDWARD GIBBON (1737-1794) ENGLISH HISTORIAN

In an autocracy, one person has his way; in an aristocracy a few people have their way; in a democracy, no one has his way.[939]

CELIA GREEN (1935-) BRITISH WRITER

If I discriminate against or criticize you, it's called religious freedom. If you return the favor, it's called persecution.[940] [Voiced as the performance character Betty Bowers, "America's Best Christian".]

DEVEN GREEN (1975-) CANADIAN COMEDIAN

The four stages of acceptance:
1. This is worthless nonsense,
2. This is an interesting, but perverse, point of view,
3. This is true, but quite unimportant,
4. I always said so.[941]

JOHN HALDANE (1892-1964) BRITISH SCIENTIST

Liberty is never found in despotism or the extremes of democracy, but in moderate governments.[942]

ALEXANDER HAMILTON (C.1755-1804) AMERICAN STATESMAN

Justice is the end of government. It is the end of civil society. It ever has been, and ever will be, pursued, until it be obtained, or until liberty be lost in the pursuit.[943]

ALEXANDER HAMILTON (C.1755-1804), JAMES MADISON (1751-1836), AND JOHN JAY (1745-1829) AMERICAN STATESMEN

Liberty lies in the hearts of men and women; when it dies there, no constitution, no law, no court can save it.[944]

LEARNED HAND (1872-1961) AMERICAN JURIST

Nowhere has democracy ever worked well without a great measure of local self-government, providing a school of political training for the people at large as much as for their future leaders.[945]

FRIEDRICH HAYEK (1899-1992) AUSTRIAN-BRITISH PHILOSOPHER

The mountebank takes advantage of this impatience in the human mind ... to tell his hearers; "This is what you are in search of; you now see and feel the truth, and need give yourself no further trouble about the mysteries of nature or the imperfection

of your own faculties."[946] [Commenting on those who exploit the human desire for meaning by extolling attractive but misleading fictions, mostly for personal gain.]

WILLIAM HAZLITT (1778-1830) ENGLISH WRITER

The garb of religion is the best cloak for power.[947]

WILLIAM HAZLITT (1778-1830) ENGLISH WRITER

Democracy is a poor system; the only thing that can be said for it is that it's eight times as good as any other method.[948]

ROBERT HEINLEIN (1907-1988) AMERICAN WRITER

It is a truism that almost any sect, cult or religion will legislate its creed into law if it acquires the political power to do so, and will follow it by suppressing opposition, subverting all education to seize the early minds of the young, and by killing, locking up, or driving underground all heretics.[949]

ROBERT HEINLEIN (1907-1988) AMERICAN WRITER

Are you a communist?
No, I am an anti-fascist.
For a long time?
Since I have understood fascism.[950]

ERNEST HEMINGWAY (1899-1961) AMERICAN WRITER

It is not a little surprising that Christianity, whose chief excellence consists of softening the human heart, in cherishing and improving its finer feelings, should encourage a practice so totally repugnant to the first impression of right and wrong. What adds to the wonder is that this abominable practice has been introduced in the most enlightened ages.[951] [Referring to the practice of slavery.]

PATRICK HENRY (1736-1799) AMERICAN STATESMAN

If one were to offer men to choose out of all the customs in the world such as seemed to them the best, they would examine the whole number, and end by preferring their own; so convinced are they that their own usages far surpass those of all others.[952]

HERODOTUS (C.484–C.425 BCE) GREEK HISTORIAN

All religions have based morality on obedience, that is to say, on voluntary slavery.[953]

ALEXANDER HERZEN (1812–1870) RUSSIAN WRITER

Between too great Liberty and too much Authority, 'tis hard to pass between the points of both unwounded.[954]

THOMAS HOBBES (1588–1679) ENGLISH PHILOSOPHER

Absolute faith corrupts as absolutely as absolute power.[955]

ERIC HOFFER (1922–1983) AMERICAN PHILOSOPHER

The all-seeing eye of a totalitarian regime is usually the watchful eye of the next-door neighbor.[956]

ERIC HOFFER (1922–1983) AMERICAN PHILOSOPHER

You measure democracy by the freedom it gives its dissidents, not the freedom it gives its assimilated conformists.[957]

ABBIE HOFFMAN (1936–1989) AMERICAN ACTIVIST

Men are idolaters, and want something to look at and kiss and hug, or throw themselves down before; they always did, they always will; and if you don't make it of wood, you must make it of words.[958]

OLIVER WENDELL HOLMES, SR. (1809–1894) AMERICAN PHYSICIAN

Democracy is not free. It comes with a price to be vigilant and hold people to account. The method that we have had for that is journalism.[959]

ROGER HOLMES (C.1969-) CANADIAN PUBLISHER

[We will ensure that] in all the book-sellers' shops and in the printing presses ... no new work shall be printed, and especially that no libelous books shall be sold, either publicly or privately; and all the rulers will make a law that if anyone wishes to publish, sell or print any new work, such works must first be revised and reviewed by certain approved and learned men.[960] [One of many such efforts to stamp out writings that impugned Catholic doctrine.]

HOLY ROMAN EMPIRE (1523)

Democracy as a way of life can flourish only when differences of opinion can be negotiated by free, critical discussion.[961]

SIDNEY HOOK (1902-1989) AMERICAN PHILOSOPHER

There is no room in a totalitarian culture for the scientific approach, with its critical probing of alternatives, when questions arise concerning the social ends and values which guide major national policies.[962]

SIDNEY HOOK (1902-1989) AMERICAN PHILOSOPHER

The problem to be considered and solved when the First Amendment was proposed was not one of hazy or comparative insignificance, but was one of blunt and stark reality, which had perplexed and plagued the nations of Western Civilization for some 14 centuries, and during that long period, the union of Church and State in the government of man had produced neither peace on earth, nor good will to man.[963]

HORACE MANN LEAGUE V BOARD (1966) COURT OF APPEALS OF MARYLAND

The more the concept of reason becomes emasculated, the more easily it lends itself to ideological manipulation and to propagation of even the most blatant lies.[964]

MAX HORKHEIMER (1895–1973) GERMAN PHILOSOPHER

The ratio of the diameter and circumference is as five-fourths to four.[965] [A convoluted proposal to legislate the "revealed" value of pi as 3.2, instead of its mathematically correct value of 3.14159 ...]

HOUSE BILL 246 (1897) INDIANA GENERAL ASSEMBLY

In all ages of the world, priests have been enemies to liberty.[966]

DAVID HUME (1711–1776) BRITISH PHILOSOPHER

Where men are the most sure and arrogant, they are commonly most mistaken.[967]

DAVID HUME (1711–1776) BRITISH PHILOSOPHER

The nature of power is such that even those who have not sought it, but have had it forced upon them, tend to acquire a taste for more.[968]

ALDOUS HUXLEY (1894–1963) BRITISH WRITER

You should never wear your best trousers when you go out to fight for freedom and truth.[969]

HENRIK IBSEN (1828–1906) NORWEGIAN PLAYWRIGHT

Error held as truth has much the effect of truth. In politics and religion this fact upsets many confident predictions.[970]

GEORGE ILES (1852–1942) CANADIAN WRITER

[Some] have said and still say, that the Ten Commandments are the foundation of all law. Nothing could be more absurd ... all

that man has accomplished for the benefit of man since the close of the Dark Ages has been done in spite of the Old Testament.[971]

ROBERT GREEN INGERSOLL (1833-1899) AMERICAN LAWYER

Nothing has so alienated nations, nothing has so destroyed the natural justice of mankind, as what has been known as religion. The idea that all men must worship the same God, believe the same dogmas, has for thousands of years plucked with bloody hands the flower of pity from the human heart.[972]

ROBERT GREEN INGERSOLL (1833-1899) AMERICAN LAWYER

I have no fear that the result of our experiment will be that men may be trusted to govern themselves without a master.[973]

THOMAS JEFFERSON (1743-1826) AMERICAN PRESIDENT

Wherever the people are well informed, they can be trusted with their own government.[974]

THOMAS JEFFERSON (1743-1826) AMERICAN PRESIDENT

I agree ... that a professorship of theology should have no place in our institution.[975] [Commenting on the curriculum at the University of Virginia, which Jefferson co-founded.]

THOMAS JEFFERSON (1743-1826) AMERICAN PRESIDENT

If a nation expects to be ignorant and free, in a state of civilization, it expects what never was and never will be.[976]

THOMAS JEFFERSON (1743-1826) AMERICAN PRESIDENT

No experiment can be more interesting than that we are now trying, and which we trust will end in establishing the fact, that man may be governed by reason and truth.[977]

THOMAS JEFFERSON (1743-1826) AMERICAN PRESIDENT

I confess that I do not see what good it does to fulminate against the English tyranny while the Roman [Catholic] tyranny occupies the palace of the soul.[978]

JAMES JOYCE (1882-1941) IRISH WRITER

No nation keeps its word.[979]

CARL JUNG (1875-1961) SWISS PSYCHIATRIST

I freely admit that the recollection of David Hume interrupted my dogmatic slumber and inspired an entirely new direction for my philosophical research.[980]

IMMANUEL KANT (1724-1804) GERMAN PHILOSOPHER

While we shall negotiate freely, we shall not negotiate freedom.[981]

JOHN KENNEDY (1917-1963) AMERICAN PRESIDENT

I think this is the most extraordinary collection of talent, of human knowledge, that has ever been gathered together at the White House, with the possible exception of when Thomas Jefferson dined alone.[982] [Remarks at a White House dinner honoring all living Nobel Prize winners.]

JOHN KENNEDY (1917-1963) AMERICAN PRESIDENT

I believe in an America where the separation of church and state is absolute; where no Catholic prelate would tell the president (should he be Catholic) how to act, and no Protestant minister would tell his parishioners for whom to vote; where no church or church school is granted any public funds or political preference; and where no man is denied public office merely because his religion differs from the President who might appoint him or the people who might elect him.[983]

JOHN KENNEDY (1917-1963) AMERICAN PRESIDENT

Free Russia is born and none will succeed in wresting the liberty from the people.[984] [On the value of political promises.]

ALEXANDER KERENSKY (1881–1970) RUSSIAN REVOLUTIONIST

Royal authority rarely is handed over voluntarily.[985]

IBN KHALDUN (1332–1406) TUNISIAN HISTORIAN

The dream of our American democracy [is] a dream yet unfulfilled.[986]

MARTIN LUTHER KING (1929–1968) AMERICAN ACTIVIST

Freedom is never voluntarily given by the oppressor; it must be demanded by the oppressed.[987]

MARTIN LUTHER KING (1929–1968) AMERICAN ACTIVIST

The ideal for a well-functioning democratic state is like the ideal for a gentleman's well-cut suit - it is not noticed.[988]

ARTHUR KOESTLER (1905–1983) HUNGARIAN WRITER

You are the nation who, rather than ruling by the Shariah of Allah in its Constitution and Laws, choose to invent your own laws as you will and desire. You separate religion from your policies, contradicting the pure nature which affirms Absolute Authority to the Lord and your Creator.[989] [Articulating, perhaps unwittingly, the most compelling argument for the separation of church from state.]

OSAMA BIN LADEN (1957–2011) SAUDI MILITANT

Humanity has won its battle. Liberty now has a country.[990] [Commenting on the outcome of the American Revolutionary War.]

MARQUIS DE LAFAYETTE (1757–1834) FRENCH MILITARY OFFICER

*I am the fellow citizen of every thinking man;
Truth is my native land.*[991]

ALPHONSE DE LAMARTINE (1790–1869) FRENCH WRITER

Even the weakest disputant is made so conceited by what he calls religion, as to think himself wiser than the wisest who thinks differently from him.[992]

WALTER SAVAGE LANDOR (1775–1864) ENGLISH WRITER

Is it not strange that the descendants of those pilgrim fathers who crossed the Atlantic to preserve their own freedom of opinion, have always proved themselves intolerant of the spiritual liberty of others.[993]

ROBERT E. LEE (1757–1834) AMERICAN MILITARY OFFICER

Under a tyranny, most friends are a liability. One quarter of them turn "reasonable" and become your enemies, one quarter are afraid to speak, and one quarter are killed and you die with them. But the blessed final quarter keep you alive.[994]

SINCLAIR LEWIS (1885–1951) AMERICAN WRITER

Four score and seven years ago our fathers brought forth on this continent, a new nation, conceived in Liberty, and dedicated to the proposition that all men are created equal.[995] [A proposition still not universally held.]

ABRAHAM LINCOLN (1809–1865) AMERICAN PRESIDENT

Totalitarian states, whether of the fascist or the communist persuasion, are more than superficially alike as dictatorships, in the suppression of dissent, and in operating planned and directed economies. They are profoundly alike.[996]

WALTER LIPPMANN (1889–1974) AMERICAN WRITER

Consent is absolutely necessary to the very being and subsistence of our government and without which our peace and religion cannot possibly be in any way secured.[997]

JOHN LOCKE (1632–1704) ENGLISH PHILOSOPHER

The end of law is not to abolish or restrain, but to preserve and enlarge freedom.[998]

JOHN LOCKE (1632–1704) ENGLISH PHILOSOPHER

Wherever law ends, tyranny begins.[999]

JOHN LOCKE (1632–1704) ENGLISH PHILOSOPHER

Freedom only for the supporters of the government, only for the members of the Party ... is no freedom at all. Freedom is always the freedom of dissenters.[1000]

ROSA LUXEMBURG (1871–1919) GERMAN PHILOSOPHER

Men must be kindly treated or utterly crushed.[1001] [Advice offered to, and frequently accepted by, despots beyond number.]

NICCOLÒ MACHIAVELLI (1469–1527) ITALIAN POLITICAL PHILOSOPHER

During almost fifteen centuries has the legal establishment of Christianity been on trial. What have been its fruits? More or less in all places, pride and indolence in the clergy, ignorance and servility in the laity, in both, superstition, bigotry and persecution.[1002]

JAMES MADISON (1751–1836) AMERICAN PRESIDENT

What influence in fact have ecclesiastical establishments had on civil society? In some instances they have been seen to erect a spiritual tyranny on the ruins of civil authority; in many instances they have been seen upholding the thrones of polit-

ical tyranny: in no instance have they been seen the guardians of the liberties of the people.[1003]

JAMES MADISON (1751-1836) AMERICAN PRESIDENT

Every nation has the government it deserves.[1004]

JOSEPH DE MAISTRE (1753-1821) FRENCH PHILOSOPHER

DEMOCRACY

If you're not ready to die for it, take the word "freedom" out of your vocabulary.[1005]

MALCOLM X (1925-1965) AMERICAN ACTIVIST

Equality before the law means the right to participate in the making of the laws by which one is governed.[1006]

NELSON MANDELA (1918-2013) SOUTH AFRICAN ACTIVIST

Without democracy there cannot be peace.[1007]

NELSON MANDELA (1918-2013) SOUTH AFRICAN ACTIVIST

There is no place in a democracy for any community or section of a community to impose its will at the expense of the fundamental rights of any other citizen.[1008]

NELSON MANDELA (1918-2013) SOUTH AFRICAN ACTIVIST

A critical, independent and investigative press is the lifeblood of any democracy.[1009]

NELSON MANDELA (1918-2013) SOUTH AFRICAN ACTIVIST

It is a strange fact that freedom and equality, the two basic ideas of democracy, are to some extent contradictory.[1010]

THOMAS MANN (1875-1955) GERMAN WRITER

The freedom of the press is one of the great bulwarks of liberty, and can never be restrained but by despotic governments.[1011]

GEORGE MASON (1725-1792) VIRGINIA STATESMAN

If any person shall presume willfully to blaspheme the holy name of God, Father, Son, or Holy Ghost; either by denying, cursing or reproaching the true God; his creation or government of the world: or by denying, cursing, or reproaching the holy word of God; every one so offending shall be punished by imprisonment, not exceeding six months, and until they find sureties for good behavior; by sitting in pillory; by whipping; boring through the tongue with a red hot iron; or sitting upon the gallows with a rope about their neck.[1012] [Religion's past idea of free speech - see next entry.]

MASSACHUSETTS COLONY (1742) BLASPHEMY LAW

Whoever willfully blasphemes the holy name of God by denying, cursing or contumeliously reproaching God, his creation, government or final judging of the world, or by cursing or contumeliously reproaching Jesus Christ or the Holy Ghost, or by cursing or contumeliously reproaching or exposing to contempt and ridicule the holy word of God contained in the holy scriptures, shall be punished by imprisonment in jail for not more than one year or by a fine of not more than three hundred dollars, and may also be bound to good behavior.[1013] [Religion's present idea of free speech - see previous entry.]

U.S. STATE OF MASSACHUSETTS (2019) BLASPHEMY LAW

The one pervading evil of democracy is the tyranny of the ma-

jority, or rather of that party, not always the majority, that succeeds, by force or fraud, in carrying elections.[1014]

THOMAS ERSKINE MAY (1815-1886) BRITISH PARLIAMENTARIAN

Disbelief in God or gods has spread always in exact proportion to the growth of knowledge and of freedom to express one's beliefs.[1015]

JOSEPH MCCABE (1867-1955) ENGLISH WRITER

Under democracy one party always devotes its chief energies [in] trying to prove that the other party is unfit to rule - and both commonly succeed, and are right.[1016]

H. L. MENCKEN (1880-1956) AMERICAN JOURNALIST

The odds are on the man who is, intrinsically, the most devious and mediocre - the man who can most adeptly disperse the notion that his mind is a virtual vacuum.[1017] [Commenting on the qualifications of sundry political candidates.]

H. L. MENCKEN (1880-1956) AMERICAN JOURNALIST

On some great and glorious day the plain folks of the land will reach their heart's desire at last, and the White House will be adorned by a downright moron.[1018]

H. L. MENCKEN (1880-1956) AMERICAN JOURNALIST

The Freedom Bell in Berlin is, like the Liberty Bell in Philadelphia, a symbol which reminds us that freedom does not come about by itself. It must be struggled for and then defended anew every day of our lives.[1019]

ANGELA MERKEL (1954-) GERMAN CHANCELLOR

The despotism of custom is everywhere the standing hindrance to human advancement.[1020]

JOHN STUART MILL (1806–1873) BRITISH PHILOSOPHER

If all mankind minus one, were of one opinion, and only one person were of the contrary opinion, mankind would be no more justified in silencing that one person, than he, if he had the power, would be justified in silencing mankind.[1021]

JOHN STUART MILL (1806–1873) BRITISH PHILOSOPHER

Let us suppose ... that the government is entirely at one with the people, and never thinks of exerting any power of coercion unless in agreement with what it conceives to be their voice. But I deny the right of the people to exercise such coercion, either by themselves or by their government. The power itself is illegitimate. The best government has no more title to it than the worst.[1022]

JOHN STUART MILL (1806–1873) BRITISH PHILOSOPHER

A party of order or stability, and a party of progress or reform, are both necessary elements of a healthy state of political life.[1023] [Reflecting on the positive friction between right- and left-leaning parties.]

JOHN STUART MILL (1806–1873) BRITISH PHILOSOPHER

The power of kings and magistrates is nothing else but what is only derivative, transferred and committed to them in trust from the people, to the common good of them all, in whom the power yet remains fundamentally and cannot be taken from them.[1024] [A pre-Enlightenment interpretation of human rights.]

JOHN MILTON (1608–1674) ENGLISH POET

None can love freedom heartily, but good men; the rest love not freedom, but license, which never hath more scope or more indulgence than under tyranny.[1025]

JOHN MILTON (1608-1674) ENGLISH POET

It is not material, in a free country, how men reason; it is enough that they do reason.[1026]

MONTESQUIEU (1689-1755) FRENCH PHILOSOPHER

The zeal for the advancement of religion differs from a due attachment to it; in order to love it and fulfil its behests, it is not necessary to hate and persecute those who are opposed to it.[1027]

MONTESQUIEU (1689-1755) FRENCH PHILOSOPHER

When the legislative and executive powers are united in the same person, or in the same body of magistrates, there can be no liberty. Again, there is no liberty, if the judiciary power be not separated from the legislative and executive.[1028]

MONTESQUIEU (1689-1755) FRENCH PHILOSOPHER

Faith, fanatic faith, once wedded fast,
To some dear falsehood, hugs it to the last.[1029]

THOMAS MOORE (1779-1852) IRISH POET

The great object is to keep the minds of the young as open as possible in the matter of religion.[1030]

JOHN MORLEY (1838-1923) BRITISH STATESMAN

We all know nations that can be identified by the flight of writers from their shores. These are regimes whose fear of unmonitored writing is justified because truth is trouble.[1031]

TONI MORRISON (1931-) AMERICAN WRITER

One of the pitfalls into which freedom can fall is in our thinking that any national or world emergency justifies putting curbs on freedom.[1032]

EDWARD R. MURROW (1908-1965) AMERICAN JOURNALIST

Only a free and unrestrained press can effectively expose deception in government.[1033]

NEW YORK TIMES V UNITED STATES (1971) UNITED STATES SUPREME COURT

In the First Amendment, the Founding Fathers gave the free press the protection it must have to fulfill its essential role in our democracy. The press was to serve the governed, not the governors.[1034]

NEW YORK TIMES V UNITED STATES (1971) UNITED STATES SUPREME COURT

People are right when they say that sport is not always good for your health.[1035] [Quipping on his arrest, as he emerged from his home to go jogging, for participating in unauthorized protests.]

ALEXEI NAVALNY (1976-) RUSSIAN ACTIVIST

Both clerical absolutism and orthodox Protestantism's principle of the divine right of kings had to be challenged before political democracy could arise.[1036]

RHEINHOLD NIEBUHR (1892-1971) AMERICAN THEOLOGIAN

Man's capacity for justice makes democracy possible, but man's inclination to injustice makes democracy necessary.[1037]

RHEINHOLD NIEBUHR (1892-1971) AMERICAN THEOLOGIAN

The struggles of our century will seem to have been mere indigestion compared with what is to come.[1038] [Predicting, in the nineteenth century, the atrocities that would arise from communism in the twentieth century.]

FRIEDRICH NIETZSCHE (1844-C.1900) GERMAN PHILOSOPHER

Democracies don't stop just with elections; they also depend on strong institutions and a vibrant civil society, and open political space, and tolerance of people who are different than you.[1039]

BARACK OBAMA (1961-) AMERICAN PRESIDENT

The greatest dangers to liberty lurk in insidious encroachment by men of zeal, well meaning but without understanding.[1040]

OLMSTEAD V UNITED STATES (1928) UNITED STATES SUPREME COURT

I do not agree with your last article, and I am canceling my wife's subscription.[1041] [Husband to magazine publisher, in response to an article on women's rights.]

ANONYMOUS

The purpose of Newspeak was not only to provide a medium of expression for the worldview and mental habits proper to the citizenry, but to make all other modes of thought impossible.[1042] [Commenting on the control of communication and thought within totalitarian regimes.]

GEORGE ORWELL (1903-1950) BRITISH WRITER

There is something wrong with a regime that requires a pyramid of corpses every few years.[1043] [Commenting on totalitarian rule, both left- and right-leaning.]

GEORGE ORWELL (1903-1950) BRITISH WRITER

Big brother is watching you.[1044] [The promise - and threat - of all despotic regimes.]

GEORGE ORWELL (1903-1950) BRITISH WRITER

War is Peace. Freedom is Slavery. Ignorance is Strength.[1045]

GEORGE ORWELL (1903-1950) BRITISH WRITER

All hereditary government is in its nature tyranny.[1046]

THOMAS PAINE (1737–1809) ENGLISH ACTIVIST

[Voltaire's] forte lay in exposing and ridiculing the superstitions which priestcraft, united with statecraft, had interwoven with governments.[1047]

THOMAS PAINE (1737–1809) ENGLISH ACTIVIST

Political despotism and supernatural religion have done more to render the human race vicious and depraved than all other causes conjointly combined.[1048]

ELIHU PALMER (1764–1806) AMERICAN WRITER

So long as the subjection of women endures and is confirmed by law and custom, so long will the race be injured and degraded, and women be victimised.[1049]

CHRISTABEL PANKHURST (1880–1958) BRITISH ACTIVIST

We are here, not because we are law-breakers; we are here in our efforts to be law-makers.[1050]

EMMELINE PANKHURST (1858–1928) BRITISH ACTIVIST

Presidential Emergency Measure Number 1 prohibits any act that denies, opposes, distorts, or slanders the Constitution of the Republic of Korea, or petitions to amend or abolish the Constitution. Any person who violates this measure may be arrested, detained, seized, or sentenced to not more than 15 years in prison without a warrant from a judge, or may be executed by order of a military tribunal.[1051] [Dislike democracy? Amend the constitution to extend your term of office, then pass draconian measures to bar further constitutional reform.]

PARK CHUNG-HEE (1917–1979) SOUTH KOREAN PRESIDENT

Xi [Jinping's] tolerance for independent voices and criticism is absolutely zero. It's non-existent.[1052] [On the tendency of rulers, long-inured to despotic regimes, to sustain tyranny.]

MINXIN PEI (1957-) CHINESE SCHOLAR

[Our] administration favours the many instead of the few: this is why it is called a democracy.[1053]

PERICLES (C.495-429 BCE) ATHENIAN STATESMAN

We regard him who takes no part in the state not as unambitious but as useless.[1054]

PERICLES (C.495-429 BCE) ATHENIAN STATESMAN

A man who is always ready to believe what is told him will never do well.[1055]

GAIUS PETRONIUS (C.27-66 CE) ROMAN COURTIER

If there is anything that will not bear free thought, let it crack.[1056]

WENDELL PHILLIPS (1811-1884) AMERICAN ABOLITIONIST

Eternal vigilance is the price of liberty - power is ever stealing from the many to the few.[1057]

WENDELL PHILLIPS (1811-1884) AMERICAN ABOLITIONIST

Democracy will never be supplanted by a republic of experts - and that is a very good thing.[1058]

THOMAS PIKETTY (1971-) FRENCH ECONOMIST

The checks and balances in a democracy ... were invented in explicit recognition of the fact that human leaders will always be tempted to arrogate power to themselves.[1059]

STEVEN PINKER (1954-) AMERICAN COGNITIVE PSYCHOLOGIST

Let there be no question of fostering under this name of Christian Democracy any intention of diminishing the spirit of obedience, or of withdrawing people from their lawful rulers. Both the natural and the Christian law command us to revere those who in their various grades are shown above us in the State, and to submit ourselves to their just commands.[1060] [Commanding, by divine mandate, the subjugation of people to civil authority.]

POPE LEO XIII (1810–1903)

There is another kind of control that the State would soon begin to exercise if it were responsible for maintaining its artists. It would remind them that he who pays the piper calls the tune. It would begin to dictate what kind of art these State servants must produce.[1061]

J. B. PRIESTLEY (1894–1984) ENGLISH WRITER

Parents may be free to become martyrs themselves. But it does not follow they are free, in identical circumstances, to make martyrs of their children before they have reached the age of full and legal discretion when they can make that choice for themselves.[1062] [Court decision limiting the rights of parents to use children to distribute religious materials.]

PRINCE V MASSACHUSETTS (1949) UNITED STATES SUPREME COURT

Atheists would not take sequestered school children and force them to recite the daily mantra that there is no god, because atheists aren't the ones imposing religious views onto other people's children, and thus denying their rights.[1063]

ARON RA (1962–) AMERICAN ACTIVIST

Wisdom does not enter into a malevolent mind, and science without conscience is but the ruin of the spirit.[1064]

FRANÇOIS RABELAIS (C.1490–1553) FRENCH WRITER

It took centuries of intellectual, philosophical development to achieve political freedom. It was a long struggle, stretching from Aristotle to John Locke to the Founding Fathers. The system they established was not based on unlimited majority rule, but on its opposite: on individual rights, which were not to be alienated by majority vote or minority plotting.[1065]

AYN RAND (1905-1982) AMERICAN WRITER

The essence of religion [is] man's surrender to god.[1066] *The essence of atheism is the freedom of the individual.*[1067] *The essence of democracy is respect for difference of opinion and acceptance of the decision by a majority of free and fluid votes.*[1068]

GOPARAJU RAMACHANDRA RAO, AKA GORA (1902-1975) INDIAN ACTIVIST

Senator, when you took your oath of office, you placed your hand on the Bible and swore to uphold the Constitution. You didn't place your hand on the Constitution and swear to uphold the Bible.[1069] [In response to a Senator who, citing religious beliefs, championed a Maryland Senate Bill to disallow same-sex marriage.]

JAMIE RASKIN (1962-) AMERICAN POLITICIAN

If ... we had an open and thoughtful exchange between Shariah scholars and American legal scholars and judges, such a dialogue could be of enormous value ... in helping Muslims to interpret Sharia in the modern Western context.[1070] [A thinly-veiled attempt to raze the barrier between church and state by introducing a chimerical deity's barbaric laws into modern democracies.]

FEISAL ABDUL RAUF (1948-) EGYPTIAN IMAM

Justice is the first virtue of social institutions, as truth is of systems of thought.[1071]

JOHN RAWLS (1921-2002) AMERICAN PHILOSOPHER

The suppression of liberty is always likely to be irrational.[1072]

JOHN RAWLS (1921-2002) AMERICAN PHILOSOPHER

Freedom is never more than one generation away from extinction. It has to be fought for and defended by each generation.[1073]

RONALD REAGAN (1911-2004) AMERICAN PRESIDENT

The silent majority is democracy's worst enemy.[1074]

JAMES ROBERTSON (1930-2019) AMERICAN HISTORIAN

Politics supersedes principles.[1075] [A discomforting but too often true observation.]

JAMES ROBERTSON (1930-2019) AMERICAN HISTORIAN

How can ... public opinion be anything but a reaction to propaganda unless there is curiosity enough in each home to keep constant watch over local, state, national and international affairs?[1076]

ELEANOR ROOSEVELT (1884-1962) AMERICAN ACTIVIST

I do not want church groups controlling the schools of our country. They must remain free.[1077]

ELEANOR ROOSEVELT (1884-1962) AMERICAN ACTIVIST

The truth is found when men are free to pursue it.[1078]

FRANKLIN D. ROOSEVELT (1882-1945) AMERICAN PRESIDENT

The demand for a statement of a candidate's religious belief can have no meaning except that there may be discrimination for or against him because of that belief. Discrimination against the holder of one faith means retaliatory discrimination against men of other faiths. The inevitable result of entering upon such a practice would be an abandonment of our real freedom of conscience and a reversion to the dreadful conditions of religious dissension which in so many lands have proved fatal to true liberty, to true religion, and to all advance in civilization.[1079]

THEODORE ROOSEVELT (1858-1919) AMERICAN PRESIDENT

It is a wicked thing either to support or to oppose a man because of the creed he professes.[1080]

THEODORE ROOSEVELT (1858-1919) AMERICAN PRESIDENT

Man is born free; and everywhere he is in chains.[1081]

JEAN-JACQUES ROUSSEAU (1712-1778) FRENCH PHILOSOPHER

The right of conquest has no foundation other than the right of the strongest.[1082]

JEAN-JACQUES ROUSSEAU (1712-1778) FRENCH PHILOSOPHER

But I am mistaken in speaking of a Christian republic; the terms are mutually exclusive.[1083]

JEAN-JACQUES ROUSSEAU (1712-1778) FRENCH PHILOSOPHER

A political party striving for power cannot be the means to the attainment of freedom.[1084]

MANABENDRA NATH ROY (1887-1954) INDIAN ACTIVIST

To me there is nothing more important in a democracy than free speech and debate. We should debate everything, we should talk about everything, we should engage ideas we aren't comfortable with, and we should let the best ideas win, that's how a healthy society based on rational ideas and a secular government should work.[1085]

DAVE RUBIN (1976-) AMERICAN COMMENTATOR

Two things form the bedrock of any open society - freedom of expression and rule of law. If you don't have those things, you don't have a free country.[1086]

SALMAN RUSHDIE (1947-) BRITISH WRITER

Man is a rational animal - so at least I have been told.[1087]

BERTRAND RUSSELL (1872-1970) BRITISH PHILOSOPHER

The fundamental cause of the trouble is that in the modern world the stupid are cocksure while the intelligent are full of doubt.[1088] [Referring to the rise of Nazi Germany and, equally, to any absolutist ideology.]

BERTRAND RUSSELL (1872-1970) BRITISH PHILOSOPHER

Which shall rule - wealth or man; which shall lead - money or intellect; who shall fill public stations - educated and patriotic freemen, or the feudal serfs of corporate capital.[1089]

EDWARD RYAN (1810-1880) AMERICAN JURIST

Science is far from a perfect instrument of knowledge. It's just the best one we have. In this respect, as in many others, it's like democracy.[1090]

CARL SAGAN (1934-1996) AMERICAN PHYSICIST

[A nation] should seek to influence mankind, not by heavy artillery, but by good example and wise counsel. It should see its highest glory, not in battles won, but in wars prevented.[1091]

CARL SCHURZ (1829-1906) AMERICAN STATESMAN

I confidently trust that the American people will prove themselves ... too wise not to detect the false pride or the dangerous ambitions or the selfish schemes which so often hide themselves under that deceptive cry of mock patriotism: "Our country, right or wrong!" They will not fail to recognize that our dignity, our free institutions and the peace and welfare of this and coming generations of Americans will be secure only as we cling to the watchword of true patriotism: "Our country - when right to be kept right; when wrong to be put right".[1092]

CARL SCHURZ (1829-1906) AMERICAN STATESMAN

Liberty cannot be gained for nothing.[1093]

SENECA THE YOUNGER (C.4 BCE-65 CE) ROMAN PHILOSOPHER

The crime of inquiry is one which religion never has forgiven.[1094]

PERCY BYSSHE SHELLEY (1792-1822) ENGLISH POET

I believe Gandhi is the only person who knew about real democracy - not democracy as the right to go and buy what you want, but democracy as the responsibility to be accountable to everyone around you.[1095]

VANDANA SHIVA (1952-) INDIAN PHILOSOPHER

What is man?
The point I want to make now is that all attempts to answer that question before 1859 are worthless, and we will be better off if we ignore them completely.[1096] [A reference to the year of publication of Darwin's *Origin of Species* and to the creation myths it displaced.]

GEORGE GAYLORD SIMPSON (1902-1984) AMERICAN PALEONTOLOGIST

Universal poverty establishes universal equality.[1097] [Noting an acerbic but historical precondition for equality.]

ADAM SMITH (1723-1790) SCOTTISH PHILOSOPHER

When we accept the statements of demagogues because we are too lazy to think about them, and test them, we can blame only ourselves for subsequent events.[1098]

MARGARET CHASE SMITH (1897–1995) AMERICAN STATESWOMAN

The right to criticize;
The right to hold unpopular beliefs;
The right to protest;
The right of independent thought.[1099]
[Stating the basic principles of American, or any, democracy.]

MARGARET CHASE SMITH (1897–1995) AMERICAN STATESWOMAN

Beware the one-party state.[1100]

TIMOTHY SNYDER (1969–) AMERICAN HISTORIAN

Do not counsel what is most pleasant, but what is best.[1101]

SOLON (C.638–C.560 BCE) ATHENIAN STATESMAN

Woe to that nation whose literature is cut short by the intrusion of force. This is not merely interference with freedom of the press but the sealing up of a nation's heart.[1102]

ALEKSANDR SOLZHENITSYN (1918–2008) RUSSIAN WRITER

Freedom has cost too much blood and agony to be relinquished at the cheap price of rhetoric.[1103]

THOMAS SOWELL (1930–) AMERICAN ECONOMIST

Every man may claim the fullest liberty to exercise his faculties compatible with the possession of like liberties by every other man.[1104]

HERBERT SPENCER (1820–1903) ENGLISH PHILOSOPHER

Modern civilization has laid the principle of freedom as its foundation, according to which man must not be a mere instrument of others, but an autonomous center of life.[1105]

ALTIERO SPINELLI (1907-1986) ITALIAN STATESMEN

Whom do you oppose? Those who forbid opposition.[1106]

LEC STANISLAW (1909-1966) POLISH POET

If the First Amendment means anything, it means that a State has no business telling a man, sitting alone in his own house, what books he may read or what films he may watch. Our whole constitutional heritage rebels at the thought of giving government the power to control men's minds.[1107]

STANLEY V GEORGIA (1969) UNITED STATES SUPREME COURT

We hold these truths to be self-evident: that all men - and women - are created equal.[1108] [Inspired from the American Declaration of Independence.]

ELIZABETH CADY STANTON (1815-1902) AMERICAN ACTIVIST

However sugarcoated and ambiguous, every form of authoritarianism must start with a belief in some group's greater right to power, whether that right is justified by sex, race, class, religion or all four.[1109]

GLORIA STEINEM (1934-) AMERICAN ACTIVIST

The great truth, I repeat, upon which our system rests, is the inferiority of the African. The enemies of our institutions ignore this truth. They set out with the assumption that the races are equal. If their premises were correct, their conclusions ... would upturn our society and lay waste to our fair country.[1110] [And so they did.]

ALEXANDER STEPHENS (1812-1883) CONFEDERATE VICE PRESIDENT

The tragedy of our day is the climate of fear in which we live, and fear breeds repression.[1111] [Denouncing a period in the United States when far-right politicians investigated "disloyal" citizens, with little regard for truth or justice.]

ADLAI STEVENSON (1900–1965) AMERICAN STATESMAN

The top one percent have the best houses, the best educations, the best doctors, and the best lifestyles, but there is one thing that money doesn't seem to have bought: an understanding that their fate is bound up with how the other ninety-nine percent live.[1112]

JOSEPH STIGLITZ (1943–) AMERICAN ECONOMIST

Is man ever a creature to be trusted with wholly irresponsible power?[1113]

HARRIET BEECHER STOWE (1811–1896) AMERICAN ABOLITIONIST

No [socialist] planning authority, however foolish, myopic, or even corrupt, could achieve quite so grandiose a misdirection and waste of the world's productive capacity as is achieved by the present phase of capitalism.[1114] [An assertion unsupported by history.]

JOHN STRACHEY (1901–1963) BRITISH POLITICIAN

Supremacy of the people is akin to freedom; between domination of a minority and the whim of a monarch the distance is small.[1115]

PUBLIUS CORNELIUS TACITUS (C.56–C.120 CE) ROMAN STATESMAN

When the state was most corrupt, laws were most abundant.[1116]

PUBLIUS CORNELIUS TACITUS (C.56–C.120 CE) ROMAN STATESMAN

1989 will be remembered for decades to come as the year when half the people of half our continent began to throw off their chains. The messages on our banners in 1979 - freedom, opportunity, family, enterprise, ownership - are now inscribed on the banners in Leipzig, Warsaw, Budapest and even Moscow.[1117] [Referring to the fall of the Berlin Wall and, more broadly, the Iron Curtain.]

MARGARET THATCHER (1925-2013) BRITISH PRIME MINISTER

When people are free to choose, they choose freedom.[1118]

MARGARET THATCHER (1925-2013) BRITISH PRIME MINISTER

The progress from an absolute to a limited monarchy, from a limited monarchy to a democracy, is a progress toward a true respect for the individual.[1119]

HENRY DAVID THOREAU (1817-1862) AMERICAN WRITER

I [do] not see why the schoolmaster should be taxed to support the priest, and not the priest the schoolmaster.[1120]

HENRY DAVID THOREAU (1817-1862) AMERICAN WRITER

In democracies public opinion has a strange power of which aristocratic nations can form no conception.[1121]

ALEXIS DE TOCQUEVILLE (1805-1859) FRENCH HISTORIAN

We repeat and again reaffirm that neither a State nor the Federal Government can constitutionally force a person "to profess a belief or disbelief in any religion." Neither can constitutionally pass laws or impose requirements which aid all religions as against nonbelievers, and neither can aid those religions based on a belief in the existence of God as against those religions founded on different beliefs.[1122]

TORCASO V WATKINS (1961) UNITED STATES SUPREME COURT

The United States of America is not in any sense founded on the Christian religion.[1123]

TREATY OF TRIPOLI (1797)

Whoever would overthrow the liberty of the nation must begin by subduing ... freedom of speech.[1124]

JOHN TRENCHARD (1662-1723) AND THOMAS GORDON (C.1691-1750) ENGLISH AND SCOTTISH COMMONWEALTHMEN

When property is taxed, the church does not hold on to it so tenaciously as when it costs nothing to keep it.[1125]

THE TRUTH SEEKER (1889) AMERICAN NEWSPAPER

There isn't any difference in totalitarian states. I don't care what you call them - you call them Nazi, Communist or Fascist, or Franco, or anything else - they are all alike.[1126]

HARRY S. TRUMAN (1884-1972) AMERICAN PRESIDENT

There was one of two things I had a right to, liberty or death.[1127] [Realizing that, despite being born into slavery, she could escape - and she did.]

HARRIET TUBMAN (C.1822-1913) AMERICAN ABOLITIONIST

Every successful revolution puts on in time the robes of the tyrant it has deposed.[1128]

BARBARA TUCHMAN (1912-1989) AMERICAN HISTORIAN

He seized the lightning from heaven and the scepter from the tyrants.[1129] [Referring to Benjamin Franklin's experiments with lightning, and to his advocacy of democracy.]

ANNE ROBERT JACQUES TURGOT (1727-1781) FRENCH ECONOMIST

Not enough of our society is trained how to understand and

interpret quantitative information. This activity is a centerpiece of science literacy to which we should all strive - the future health, wealth, and security of our democracy depend on it.[1130]

NEIL DEGRASSE TYSON (1958-) AMERICAN ASTROPHYSICIST

1. Everyone has the right to take part in the government of his country, directly or through freely chosen representatives.
2. Everyone has the right of equal access to public service in his country.
3. The will of the people shall be the basis of the authority of government; this will shall be expressed in periodic and genuine elections which shall be by universal and equal suffrage and shall be held by secret vote or by equivalent free voting procedures.[1131]

UNIVERSAL DECLARATION OF HUMAN RIGHTS (1948) ARTICLE 21

No man shall be compelled to frequent or support any religious worship, place, or ministry whatsoever, nor shall be enforced, restrained, molested, or [burdened] in his body or goods, nor shall otherwise suffer, on account of his religious opinions or belief.[1132]

VIRGINIA ASSEMBLY (1779)

It is dangerous to be right in matters where the established authorities are wrong.[1133]

VOLTAIRE (1694-1778) FRENCH PHILOSOPHER

In a civilized society freedom to offend should be protected.[1134]

NIGEL WARBURTON (1962-) BRITISH PHILOSOPHER

The establishment of our new government seemed to be the last great experiment for promoting human happiness by reasonable compact in civil society.[1135]

GEORGE WASHINGTON (1732-1799) AMERICAN PRESIDENT

Democracy dies in darkness.[1136] [On the necessity of transparency in government.]

MOTTO OF THE WASHINGTON POST (2017) AMERICAN NEWSPAPER

There is no more harmful distortion of political force than the parvenu-like braggart with power, and the vain self-reflection in the feeling of power, and in general the very worship of power.[1137]

MAX WEBER (1864-1920) GERMAN SOCIOLOGIST

Politics is made with the head, not the soul.[1138]

MAX WEBER (1864-1920) GERMAN SOCIOLOGIST

Whenever one tries to suppress doubt, there is tyranny.[1139]

SIMONE WEIL (1909-1943) FRENCH PHILOSOPHER

There is no such source and cause of strife, quarrel, fights, malignant opposition, persecution, and war, and all evil in the State, as religion. Let it once enter into our civil affairs, our government would soon be destroyed. Let it once enter into our common schools, they would be destroyed. Those who made our Constitution saw this, and used the most apt and comprehensive language in it to prevent such a catastrophe.[1140]

WEISS V DISTRICT BOARD (1890) SUPREME COURT OF WISCONSIN

[Muhammed's] own rule was unlimited autocracy, and autocratic Islam has remained. Politically, Islam was not an advance, but a retrogression from the traditional freedoms and customary laws of the desert. Its master has always been whatever man was vigorous and unscrupulous enough to snatch and hold the Caliphate, and its final law has been that of man's

will.[1141]

H. G. WELLS (1866-1946) BRITISH WRITER

You can take our lives and our freedom, but not our honor.[1142] [Speaking against the rise of the Nazi party.]

OTTO WELS (1873-1939) GERMAN POLITICIAN

Democracy is the recurrent suspicion that more than half of the people are right more than half of the time.[1143]

E. B. WHITE (1899-1985) AMERICAN ESSAYIST

If believers feel that their faith is trivialized and their true selves compromised by a society that will not give religious imperatives special weight, their problem is not that secularists are anti-democratic but that democracy is anti-absolutist.[1144]

ELLEN WILLIS (1941-2006) AMERICAN WRITER

This is surely a very great and direct moral evil in the constitution.[1145] [Arguing that the Constitution of the United States, and in particular the First Amendment, contravenes divine will by failing to acknowledge the nation's subjugation to a deity.]

JAMES R. WILLSON (1780-1853) AMERICAN THEOLOGIAN

The history of liberty is a history of the limitation of governmental power, not the increase of it.[1146]

WOODROW WILSON (1856-1924) AMERICAN PRESIDENT

No nation should seek to extend its polity over any other nation or people, but that every people should be left free to determine its own polity, its own way of development, unhindered, unthreatened, unafraid, the little along with the great and powerful.[1147]

WOODROW WILSON (1856-1924) AMERICAN PRESIDENT

A cult is a religion with no political power.[1148]

TOM WOLFE (1930-2018) AMERICAN WRITER

Do you hear the people sing?[1149] [A reference to activists in Hong Kong, who broke out in song during a protest for democracy and free elections. The song, from the musical *Les Miserables*, contains the words: *Do you hear the people sing; singing the songs of angry men; it is the music of the people; who will not be slaves again*.]

JOSHUA WONG (1996-) HONG KONG ACTIVIST

Teachers of science in the public schools should not be expected to avoid the discussion of every scientific issue on which some religion claims expertise.[1150]

WRIGHT V HOUSTON INDEPENDENT SCHOOL DISTRICT (1972) UNITED STATES DISTRICT COURT

[For] subversion of state power.[1151] [The charge that led to a 13-year prison sentence for democracy activist Qin Yongmin.]

WUHAN CITY INTERMEDIATE PEOPLE'S COURT (2018) CHINA

We believe in the treasured principle of one man, one vote. And I am that man.[1152]

PARODY OF XI JINPING (1953-) CHINESE PRESIDENT

All work of the party's news and public opinion media must reflect the will of the party, mirror the views of the party, preserve the authority of the party, preserve the unity of the party, and achieve love of the party, protection of the party, and acting for the party.[1153] [Totalitarianism redux.]

XI JINPING (1953-) CHINESE PRESIDENT

One child, one teacher, one book and one pen can change the

world.[1154] [On the role of education in cultivating and sustaining free societies, and on its power to combat extremist ideologies.]

MALALA YOUSAFZAI (1997-) PAKISTANI ACTIVIST

If you don't know history, it's as if you were born yesterday. If you were born yesterday, then any leader can tell you anything.[1155]

HOWARD ZINN (1924-2010) AMERICAN HISTORIAN

I accuse.[1156] [The title of an open letter accusing the French government of a miscarriage of justice, arguably becoming the most famous newspaper headline in history.]

ÉMILE ZOLA (1840-1902) FRENCH WRITER

The day that this country ceases to be free for irreligion it will cease to be free for religion.[1157]

ZORACH V CLAUSEN (1952) UNITED STATES SUPREME COURT

Liberty, democracy, and legality are never condescendingly bestowed gifts. All those who seek after truth and love freedom should make unremitting efforts to actually gain the rights our constitution promised to every citizen - namely, freedom of thought, freedom of speech, freedom of the press, freedom of publication, freedom of association, freedom of assembly, and freedom of demonstration.[1158]

500 BEIJING FACULTY MEMBERS (1989) PRO-DEMOCRACY STATEMENT

MISCELLANY

Enlightenment humanism possesses no central authority - no pope, prophet, or potentate. As such, there is no autocracy to impose upon the philosophy a rigid set of precepts or a uniform nomenclature. This is as it should be. One consequence of this reality, however, is that *Enlightenment humanist* is but one of several labels adopted by proponents of Enlightenment principles. Other terms include *secular humanist*, *natural humanist*, *evolutionary humanist*, *scientific humanist*, *social humanist*, or simply, in its modern context, *humanist*.

ABOUT THE AUTHOR

Critique the idea, not its proponent.
SERGE BOURGAINE

This chapter is mislabeled. Serge is not the author but the editor of this volume. The authors are those who penned or voiced the individual quotations. It is their intellect, acuity, and occasional wit that, Serge hopes, inspires the reader to further explore the tenets of Enlightenment humanism.

Of Serge himself there is little to say, as Serge Bourgaine is a pseudonym. The use of a pen name reflects no attempt at anonymity. It merely affirms that the editor is unimportant. It is the quotations and their authors who merit thought and reflection. Likewise, it is this assemblage of thinkers, writers, and luminaries who deserve whatever credit the reader wishes to bestow.

Serge

ENDNOTES

Entries that omit page numbers refer, in nearly all instances, to ebooks or similar digital content.

[1] Veenhoven 2007, not paginated.

[2] Adams 2002, not paginated.

[3] Adams 1824, p. 117.

[4] Allen 1836, p. 46.

[5] Allen 1990, p. 411.

[6] Allen 1994, p. 245.

[7] Allen 1980, timestamp 00:43:38.

[8] Bragg, Genetic Engineering, 14 January 1999, timestamp 00:23:15.

[9] Heath 1913, p. 302.

[10] Aristotle 2003a, Bekker 1102b25, p. 348. Source text: *No doubt, however, we must none the less suppose that in the soul too there is something contrary to the rational principle, resisting and opposing it.*

[11] Arnold 1903a, p. xxix. Source text: *The freethinking of one age is the common sense of the next, and the Christian world will certainly learn to transform beliefs which it now thinks to be untransformable.*

[12] Arnold 1903b, p. 57.

[13] Arnold 1903a, p. xiv.

[14] Asimov 1981, not paginated.

[15] Asimov 1983, p. 43.

[16] Asimov 1988, p. 5.

[17] Atkins 1996, paraphrased from p. 6.

[18] Atkins 2000, paraphrased from pp. 25-31.

[19] Augustine of Hippo 1892, p. 52.
[20] King James Bible Online, John 21:11.
[21] King James Bible Online, Isaiah 11:1-3.
[22] Augustine of Hippo 1874, tractate 122, p. 536.
[23] Austin 1983, p. 620.
[24] Ayala 1974, p. 700.
[25] Ayer 1952, p. 120.
[26] Bacon 2003, paraphrased from p. 114.
[27] Bacon 2003, p. 121.
[28] Bacon 2003, p. 124.
[29] Bacon 1908, pp. 233-234.
[30] Ball 2018, not paginated.
[31] Barker 2008, p. 101.
[32] Barnes 1933, lecture 1, p. 1.
[33] Batson 1993, p. 379.
[34] Beauvoir 1976, p. 242. Source text: *La représentation du monde comme le monde lui-même est l'opération des hommes; ils le décrivent du point de vue qui est le leur et qu'ils confondent avec la vérité absolue*.
[35] Bell Burnell 2010, not paginated.
[36] Berkeley 1906, p. 67.
[37] King James Bible Online, 1 Thessalonians 5:21.
[38] New International Version Bible.
[39] New International Version Bible.
[40] New International Version Bible.
[41] Bierce 1911, p. 343.
[42] Bierce 1993, not paginated.
[43] Boileau 1872, p. 288. Source text: *Aimez donc la raison: que toujours vos écrits, empruntent d'elle seule et leur lustre et leur prix*.
[44] Krucoff 1984, p. K8.
[45] Bonner 1919, p. 14.
[46] Boyer 2001, p. 320.
[47] Boyer 2001, paraphrased from p. 297.
[48] Bradbury 1979, p. 117.
[49] Bentley 1966, p. 18. Note: Per Bentley, the quoted text appears only in early versions of Brecht's play *Galileo*.
[50] Bronowski 1956, p. 18.
[51] Brooks 1933, p. 103.
[52] Buchan 1941, p. 201.

[53] Burbank 1926, p. 9.

[54] Burroughs 1900, p. 19.

[55] Bury 1913, p. 20.

[56] Calder 1970, p. 25.

[57] Calvin 1845b, paraphrased from p. 261.

[58] Campbell 1972, p. 253.

[59] Campbell 1999, p. 14.

[60] Carlin 1997, p. 227.

[61] Wood 1893, p. 190.

[62] Lear 1998, p. 148.

[63] Cathcart 2008, p. 57. Source text: *A scientist and his wife are out for a drive in the country. The wife says, "Oh, look! Those sheep have been shorn." "Yes," says the scientist. "On this side."*

[64] Chandler 1976, p. 7.

[65] Chartier 1920, p. 20. Source text: *La liberté intellectuelle, ou sagesse, c'est le doute. Cela n'est pas bien compris, communément.*

[66] Châtelet 1740, maxim 64, p. 84. Source text: *Une experience ne suffit pas pour admettre une hipothese, mais une seule suffit pour la rejetter lorsqu'elle lui est contraire.*

[67] Chesterton 1908, p. 15.

[68] Barsky 1997, p. 141.

[69] Cicero 1960, pp. 174-175. Source text: *Non enim parum cognosse, sed in parum cognito stulte et diu perseverasse turpe est, propterea quod alterum communi hominum infirmitati, alterum singulari cuiusque vitio est adtributum.*

[70] Clifford 1879a, p. 104.

[71] Clifford 1879b, p. 125-6.

[72] Cohen 1980, p. 258.

[73] Cohen 1955, p. 394.

[74] Coleridge 1895, p. 261.

[75] Collingwood 1947, maxim 1.46, paraphrased from p. 3.

[76] Colton 1829, reflection 121, pp. 83-84.

[77] Colton 1825, reflection 115, p. 69.

[78] Finocchiaro 1989, paraphrased from pp. 147-148.

[79] Finocchiaro 1989, paraphrased from pp. 287-291.

[80] Conklin 1937, p. 601.

[81] Copernicus 1566, book 1, chapter 10. Source text: *In medio uero omnium residet sol.*

[82] Copernicus 1566, Preface Letter to Pope Paul III. Translation from

Dorothy Stimson in The Gradual Acceptance of the Copernican Theory of the Universe, 1917. Source text: *Si fortasse erunt ματαιολόγοι, qui cum omnium mathematum ignari sint, tamen de illis iudicium sibi sumunt, propter aliquem locum scripturæ, male ad suum propositum detortum, ausi fuerint meum hoc institutum reprehendere ac insectari: illos nihil moror, adeo ut etiam illorum iudicium tanquam temerarium contemnam.*

83 Aitkenhead 2016, not paginated.

84 Connor 2011, not paginated.

85 Curie 1947, p. 331.

86 Cuvier 1810, p. 293. Source text: *Conduire l'esprit humain à sa noble destination, sa connaissance de la vérité; répandre des idées saines jusque dans les classes les moins élevées du peuple; soustraire les hommes à l'empire des préjugés et des passions; faire de la raison l'arbitre et le guide suprême de l'opinion publique, voilà l'objet essentiel des sciences; voilà comment elles concourent à avancer la civilisation.*

87 Darwin 2003, p. 253.

88 Dawkins 1996, timestamp 00:10:04.

89 Dawkins 2009, not paginated.

90 Dawkins 2009, not paginated.

91 Demosthenes 1930, pp. 52-53. Source text: *ὅ γὰρ βούλεται, τοῦθ' ἕκαστος καὶ οἴεται.*

92 Dennert 1903, paraphrased from p. 64. Source text: *Dieser Artikel kann die frohe Botschaft bringen dass sich ... der Darwinismus zum Sterben niederlegt. Bis er dort ganz tot ist und nicht mehr als Waffe gegen das Christentum geschwungen wird, bis dahin wird ja freilich noch viel Druckerschwärze vergeudet werden, allein es ist doch immerhin ein Anfang gemacht.*

93 Dennett 1995, p. 19.

94 Dennett 1995, p. 187.

95 Dennett 2017, p. 82.

96 Descartes 1911, principle 1, p. 219.

97 Descartes 1910, p. 44. Source text: *Ce n'est pas assez d'avoir l'esprit bon, mais le principal est de l'appliquer bien.*

98 Deschner 1996, p. 35. Source text: *Ich denke, also bin ich kein Christ.*

99 Descartes 1910, p. 19. Source text: *Je doute, donc je pense; je pense, donc je suis.*

100 Deschner 2013, p. 369. Source text: *Da das oberste Kriterium für die*

Rezeption naturwissenschaftlicher Theorien ihre Vereinbarkeit mit der Bibel war, stagnierte die Naturwissenschaft jetzt nicht nur, sondern man gab sogar preis, was man längst gewonnen hatte. Das Ansehen der Naturwissenschaft sank im selben Maße, in dem das der Bibel stieg.

101 Dewey 1929, p. 294.

102 Dewey 1910a, p. 13.

103 Dick 1996, p. 261.

104 Dickens 1998, p. 332.

105 Diderot 1821b, aphorism 31, p. 217. Source text: *Le scepticisme est donc le premier pas vers la vérité.*

106 Draper 1880, p. vi.

107 Duclaux 1896, p. 142. Source text: *On voit bien ici de quelle série de jugements sans cesse revisés se fait le progrès incontestable de la science.*

108 Duhem 1906, p. 28. Source text: *L'accord avec l'expérience est, pour une théorie physique, l'unique critérium de vérité.*

109 Dumas 1880, p. 114. Source text: *Voilà ce que c'est que la verité! C'est une dure compagne, elle est hérissée de fer; elle blesse qui elle atteint, et parfois aussi qui la dit.*

110 St. John 1972, p. 382.

111 Eddington 1929, pp. 24-25.

112 Eddington 1920, paraphrased from p. 24.

113 Einstein 1919, supplementary sheet 4. Source text: *Niemals aber kann die Wahrheit einer Theorie erwiesen werden. Denn niemals weiss man, dass auch in Zukunft keine Erfahrung bekannt werden wird, die ihren Folgerungen widerspricht.*

114 Forst de Battaglia 1970, p. 107.

115 Eiseley 1975, p. 5.

116 Eldridge 1985, p. 13.

117 Elsberry 2007, p. 252.

118 Fahringer 1991, page M06.

119 Rosenfeld 1987, not paginated.

120 Feynman 1985, p. 343.

121 Gleich 1992, p. 372.

122 Feynman 1994, p. 150.

123 Fichte 1806, paraphrased from p. 23. Source text: *Die Wissenschaft endlich setzt über alle Zeit und alle Zeitalter hinweg, indem sie die Eine, sich selber gleiche Zeit als den höheren Grund aller Zeitalter*

erfasst, und ihrer freien Betrachtung unterwirft.

[124] Fielding 1902, p. 142.

[125] Fisher 1974, p. 7.

[126] Fosdick 1941, pp. 140-141.

[127] Foucault 2006, p. 244.

[128] Foxe 1837, p. 720. Note: In the 1570 edition, the remark appears as: *Either the Pope must abolishe printyng, or hee must seke a new worlde to reigne ouer*. Foxe attributes the comment to an unnamed vicar in Croydon, England.

[129] France 1895, p. 10. Source text: *Ce qui est admirable, ce n'est pas que le champ des étoiles soit si vaste, c'est que l'homme l'ait mesuré*.

[130] Franklin 1758, not paginated.

[131] Maddox 2002, p. 61.

[132] Maddox 2002, p. 61.

[133] Maddox 2002, p. 61.

[134] Freud 1933, pp. 243-244. Source text: Keine Verkleinerung der Wissenschaft kann auch etwas an der Tatsache ändern, dass sie versucht, unserer Abhängigkeit von der realen Aussenwelt gerecht zu werden, während die Religion Illusion ist und ihre Stärke aud dem Entgegenkommen gegen unsere Triebwunschregungen bezieht.

[135] Freud 1927, p. 50. Source text: *Wender wir uns nach dieser Orientierung wieder zu den religiösen Lehren, so dürfen wir wiederholend sagen: Sie sind sämtlich Illusionen, unbeweisbar, niemand darf gezwungen, sie für wahr zu halten, an sie zu glauben*.

[136] Fry 2001, timestamp 00:12:49.

[137] Fuller 1845, p. 7.

[138] Arago 1857, p. 236.

[139] Galilei 1623, p. 25. Source text: *La filosofia è scritta in questo grandissimo libro, che continuamente ci stà aperto innanzi à gli occhi (io dico l'universo), ma non si può intendere se prima non s'impara à intender la lingua, e conoscer i caratteri, ne quali è scritto. Egli è scritto in lingua matematica*.

[140] Galilei 1967, p. 115.

[141] Gardner 1996, p. 271.

[142] Gervais 2010, not paginated.

[143] Ghosh 2008, p. 6.

[144] Bradley 1996, lines 464-466, p. 503.

[145] Goethe 1914, p. 65. Source text: *Der Mensch ist gemacht, dass man ihn das Abenteuerlichste überreden kann; das haftet aber auch gleich so fest, und wehe dem, der es wieder auskratzen und austilgen will.*

[146] Gorham 1930, p. 29.

[147] Gould 1995, p. 45.

[148] Gould 2007, not paginated.

[149] Gould 1992, p. 437.

[150] Gourmont 1923, p. 23. Source text: *Comme dit le notaire, qui en fit la dernière vente, elle est à usage de ferme, et c'est un grand progrès que là où des femmes priaient, des vaches ruminent.*

[151] Grayling 2006, not paginated.

[152] Grayling 2009, not paginated.

[153] Green 1977, p. 12.

[154] Green 1977, p. 1.

[155] Grimmelshausen 1899, p. 10. Source text: *Ich war so perfekt und vollkommen in der Unwissenheit, dass mir unmöglich war zu wissen, dass ich sogar nichts wusste.*

[156] The Hadith, Sunan Ibn Majah, 1:229.

[157] Haldane 1934, pp. vi-vii.

[158] Williamson 1992, p. 14 (quoted by A. E. Slater).

[159] Hardy 1967, p. 130.

[160] Harris 2005b, timestamp 00:45:40.

[161] Harris 2005a, p. 14.

[162] Hart 1914, p. 136.

[163] Hawking 1993, p. 134.

[164] Hawking 1988, not paginated. Source text: *Wir sind nur eine etwas fortgeschrittene Brut von Affen auf einem kleinen Planeten, der um einen höchst durchschnittlichen Stern kreist. Aber wir können das Universum verstehen, und das macht aus uns etwas sehr Besonderes.*

[165] Hawking 1998, p. 80.

[166] Onyanga-Omara 2018, not paginated.

[167] Masood 2009, p. 169.

[168] Hazlitt 1934, p. 201.

[169] Hazlitt 1902b, p. 47.

[170] Heehler 2011, p. 58.

[171] Heine 1900, p. 126. Source text: *Dort wo man Bücher verbrennt,*

verbrennt man auch am Ende Menschen.

[172] Heinlein 1966, p. 134.

[173] Herbert 1987, p. 128.

[174] Hibben 1908, p. 430.

[175] Singh 2005, p. 496.

[176] Bowen 1967, p. 536.

[177] Hitchens 2007, p. 150.

[178] Hitchens 1982, not paginated.

[179] Hobbes 2003, p. 59.

[180] Hoffer 1974, reflection 32, p. 22.

[181] Denny 1938, p. 21

[182] Doyle 1891, p. 403.

[183] Ashton 2009, paraphrased from p. 283.

[184] Hook 1950, p. 230.

[185] Hook 1950, p. 226.

[186] Horace 1942, pp. 264-265. Source text: Sapere aude; incipe.

[187] Hossenfelder 2019, timestamp 00:10:20.

[188] Laurie 2007, timestamp 00:29:25.

[189] Hubbard 1916, p. 13.

[190] Hubble 1954, p. 6.

[191] Hugo 1877, p. 236. Source text: *Il y a maintenant en France dans chaque village un flambeau allumé, le maître d'école, et une bouche qui souffle dessus, le curé.*

[192] Humboldt 1845, paraphrased from p. 5. Source text: *Die dogmatischen Ansichten der vorigen Jahrhunderte leben dann nur fort in den Vorurtheilen [sic] des Volks und in gewissen Disciplinen, die, in dem Bewusstsein ihrer Schwäche, sich gern in Dunkelheit hüllen.*

[193] Hume 1960, p. 272.

[194] Hume 2003, p. 491.

[195] Hume 2003, p. 509. Note: Hume also challenged the absolute truth of *a posteriori* reasoning, but not its affirmation of provisional truth.

[196] Huxley 1957, p. 205.

[197] Huxley 1931, p. 9.

[198] Huxley 1898, paraphrased from p. 244.

[199] Huxley 1903, p. 15.

[200] Huxley 1904, p. 41.

[201] Ingersoll 1921, p. 309.

[202] Ingersoll 1909b, p. 462.

[203] Ingersoll 1909b, p. 358.

[204] Jacoby 2018, p. 22.

[205] James 1907, p. 194.

[206] James 2003, p. 41.

[207] Jefferson 1820a, p. 1. Source text: *The most restive is that of the priests of the different religious sects who dread the advance of science as witches do the approach of daylight.*

[208] Jefferson 1820b, p. 1.

[209] Jefferson 1816b, p. 1.

[210] Jemison 2005, not paginated.

[211] Jobs 2001, not paginated.

[212] Jung 1970, pp. 117-118.

[213] Kahneman 2011, p. 62.

[214] Kahneman 2011, p. 201.

[215] Kaku 1995, p. 314.

[216] Kant 1784, paraphrased from p. 481. Source text: *Sapere Aude! Habe Muth dich deines eigenen Verstandes zu bedienen! ist also der Wahlspruch der Aufklärung.*

[217] Kaufmann 1963, p. 103.

[218] Kennedy 1962a, p. 2.

[219] Burtt 1925, p. 50.

[220] Kepler 1609, penultimate page of Introduction. Source text: *In Theologia quidem authoritatum, in Philosophia vero rationum esse momenta ponderanda.*

[221] Al-Khalili 2011, not paginated.

[222] Kierkegaard 1958, p. 115. Source text: De omnibus dubitandum est.

[223] Verbally communicated to the editor.

[224] King 2012, not paginated.

[225] Kitzmiller 2005, not paginated.

[226] Knight 1995, p. 4.

[227] Knight 1865, p. 31.

[228] Krauss 2012, p. 17.

[229] Krauthammer 2018, not paginated.

[230] Kurtz 2012, timestamp 00:22:55.

[231] La Follette 1926, pp. 2-3.

[232] Lamont 1957, p. 108.

[233] Rosenfeld 1966, p. 320.

[234] Latham 1878, p. 388. Source text: *A small overweight of knowledge is often a sore impediment to the movements of common sense.*

[235] Lavoisier 1793, Preface, p. x. Source text: *Le seul moyen de prévenir ces écarts consiste... à le mettre continuellement à l'épreuve de l'expérience; à ne conserver que les faits qui ne font que des données de la nature, & qui ne peuvent nous tromper; à ne chercher la vérité que dans l'enchaînement naturel des expériences & des observations..*

[236] Lecky 1917, pp. 351-352.

[237] Lederman 1993, p. 198.

[238] Lehrer 1965, not paginated.

[239] Lévi-Strauss 1964, p. 15. Source text: *Le savant n'est pas l'homme qui fournit les vraies réponses; c'est celui qui pose les vraies questions.*

[240] Lichtenberg 1906, aphorism F660, p. 241. Source text: *Es giebt [sic] eine Art von transzendenter Ventriloquentz [sic] wodurch Menschen können glauben gemacht werden, etwas was auf Erden gesagt ist käme vom Himmel.*

[241] Lippman 1915, p. 51.

[242] Locke 1794, p. 271.

[243] Locke 1882, p. 30.

[244] Ballou 1881, p. 433.

[245] Ballou 1886, p. 465.

[246] Locke 2003b, pp. 387-388.

[247] Luther 1959, p. 99.

[248] Luther 1957, p. 320.

[249] Luther 1914, p. 126. Source text: *Ratio ... ist die hochste hur, die der Teuffel hat.*

[250] Macaulay 1860, p. 304.

[251] Macrobius 1970, p. 409. Source text: *Plebeia ingenia magis exemplis quam ratione capiuntur.*

[252] Magnus 1967, p. 69.

[253] Maher 2005, timestamp 01:08:20.

[254] Manilius 1920, p. 125. Source text: Ratio omnia vincit.

[255] Mann 1969a, p. 57.

[256] Marlowe 1967, p. 9.

[257] Robertson 1906, p. 269.

[258] Diderot 1779, p. 667. Source text: *La raison est à l'égard du philosophe, ce que la grace est à l'égard du chretien.*

[259] Massay 2019, p. 325.

[260] Maugham 1919, paraphrased from p. 11. Source text: *The modern*

clergyman has acquired in his study of the science which I believe is called exegesis an astonishing facility for explaining things away.

[261] Maugham 1938, paraphrased from p. 50.

[262] Campbell 1882, p. 159.

[263] McCabe 1933, p. 80.

[264] McKinney 2010, p. 281.

[265] McKown 1993, p. 39.

[266] Mead 1970, p. 72. This widely quoted truism appears to be a rephrasing. The likely source text: *We must create new models for adults who can teach their children not what to learn, but how to think, and not what they should be committed to, but the value of commitment.*

[267] Medawar 1982, pp. 110-111.

[268] Medawar 1981, p. 39.

[269] Sime 1996, p. 375.

[270] Mencken 1956, aphorism 412, p. 277.

[271] Mill 1836, p. 25.

[272] Mill 2003, p. 284.

[273] Mill 1874, p. 239.

[274] Mirzakhani 2014, not paginated.

[275] Mises 1962, p. 94.

[276] Molière 1907, p. 8. Source text: *La parfaite raison fuit toute extremité et veut que l'on soit sage avec sobrieté.*

[277] Montagu 1984, pp. 8-9.

[278] Montaigne 1906, p. 314. Source text: *Qui veut guérir de l'ignorance, il faut la confesser.*

[279] Montaigne 2003, p. 295.

[280] Montessori 1912, p. 8.

[281] Monty Python 1979, timestamp 01:01:28.

[282] Morgan 1915, pp. 1-2. Source text: Je *n'avais pas besoin de cette hypothèse-là.*

[283] Morris 1978, p. 94.

[284] Murasaki 2000, p. 180.

[285] Murray 1965, p. 62.

[286] Gillispie 1980, p. 450. Source text: *Das Wissen ist der Tod der Forschung.*

[287] Newton 2003, folio 88r. Source text: *Amicus Plato amicus Aristoteles magis amica veritas.*

[288] Niebuhr 2010, p. 2.

[289] Nietzsche 1922b, aphorism 822, p. 248. Source text: *Wir haben die Kunst, damit wir nicht an der Wahrheit zu Grunde gehn.*

[290] Nietzsche 1921, aphorism 20, p. 22. Source text: *Der Glaube an die Wahrheit beginnt mit dem Zweifel an allen bis dahin geglaubten "Wahrheiten".*

[291] Nietzsche 1919b, p. 290. Source text: *"Glauben" heisst Nicht-wissen-wollen, was wahr ist.*

[292] Nye 2012, timestamp 00:01:45.

[293] Nye 2014, paraphrased from timestamp 02:03:54 through 02:07:18.

[294] Ockham 1995, p. 4.

[295] Oliver 2016, timestamp 00:14:50.

[296] Coffin 1935, not paginated.

[297] Lucas 1979, paraphrased from p. 314.

[298] Paine 1894, p. 233.

[299] Paine 1837, p. 284.

[300] Paine 1908, p. 22.

[301] Pareto 1919, p. 131. Source text: *Gli uomini, quando a loro giovi, possono prestare fede a una teoria di cui conoscono poco più che il nome; il che del rimanente è fenomeno generale in tutte le religioni.*

[302] Parfit 1989, p. 281.

[303] Parker 1863, p. 78.

[304] Burnet 1908, p. 197.

[305] Pascal 1918, p. xlv. Source text: *La raison nous commande bien plus impérieusement qu'un maître; car en désobéissant à l'un on est malheureux, et un désobéissant à l'autre on est un sot.*

[306] Kaufmann 1994, p. 522.

[307] Payne-Gaposchkin 1984, p. 233.

[308] Peirce 1901, p. 322.

[309] Peirce 1932, p. 56.

[310] Penn 1903, maxim 156, pp. 85-86.

[311] Thucydides 2003, p. 397.

[312] Pinker 2015, not paginated.

[313] Plait 2015, timestamp 00:01:05.

[314] Planck 1949, p. 325.

[315] Plato 2003a, Stephanus 38, p. 210.

[316] Plato 2003d, Stephanus 313, p. 40.

[317] Poincaré 1920, p. 137. Source text: *C'est par la logique qu'on*

démontre, c'est par l'intuition qu'on invente.

[318] Poincaré 1917, p. 167. Source text: *L'expérience est la source unique de la vérité: elle seule peut nous apprendre quelque chose de nouveau; elle seule peut nous donner la certitude.*

[319] Pope Leo XIII 1888, article 21.

[320] Pope Leo XIII 1888, article 23.

[321] Pope Paul VI 1964, article 25. Source text: *This religious submission of mind and will must be shown in a special way to the authentic magisterium of the Roman Pontiff, even when he is not speaking ex cathedra; that is, it must be shown in such a way that his supreme magisterium is acknowledged with reverence, the judgments made by him are sincerely adhered to, according to his manifest mind and will.*

[322] Popper 1965, p. 216.

[323] Ayer 1968, p. 282.

[324] Popper 2002, p. 18.

[325] Popper 2002, p. 281.

[326] Ra 2015b, question 27, not paginated.

[327] Ra 2017, not paginated.

[328] Rabi 1964, not paginated.

[329] Ramasamy 2007, not paginated.

[330] Brandon 1986, p. 165.

[331] Rao 1978, p. 48.

[332] Ratner 1936, pp. 109-110.

[333] Ratner 1936, p. 104.

[334] Reade 1887, p. 34.

[335] Meyer 2019, not paginated.

[336] Renan 1863, p. 50 of Introduction. Source text: *Aucun des miracles dont les vieilles histoires sont remplies ne s'est passé dans des conditions scientifiques. Une observation qui n'a pas été une seule fois démentie nous apprend qu'il n'arrive de miracles que dans les temps et les pays où l'on y croit, devant des personnes disposées à y croire.*

[337] Pinker 2012, p. 200.

[338] Root-Berstein 1995, p. 38.

[339] Root-Berstein 1995, p. 38.

[340] Root-Berstein 1995, p. 41.

[341] Rousseau 1912, p. 81. Source text: *Le chef-d'œuvre d'une bonne éducation est de faire un homme raisonnable: et l'on prétend*

élever un enfant par la raison! C'est commencer par la fin, c'est vouloire faire l'instrument de l'ouvrage. Si les enfants entendaient raison, ils n'auraient pas besoin d'être élevés.

[342] The Royal Society, not paginated.

[343] Royce 1934, p. 79. Source text: *Error is not a mere accident of an untrained intellect, but a necessary stage or feature or moment of the expression of the truth.*

[344] Koupelis 2007, p. 583 (margin).

[345] Ruskin 1909, paraphrased from p. 115.

[346] Russell 1953, p. 7.

[347] Russell 1950, pp. 73-74.

[348] Russell 1958, p. 161.

[349] Rutherford 2017, not paginated.

[350] Cohen 1985, pp. 105-106.

[351] Bernal 2017, paraphrased from p. 9.

[352] Sagan 1990, not paginated.

[353] Sagan 1996, not paginated.

[354] Sagan 1996, not paginated.

[355] Sagan 1996, not paginated.

[356] Sagan 2000, timestamp 00:01:25.

[357] Sagan 1980, p. 333.

[358] Sagan 1980, p. 218.

[359] Santayana 1929, p. 69.

[360] Santayana 1922, p. 246.

[361] Schiller 1878, p. 131. Source text: *Schwer ist es, aus dem Geschrei erhitzter Parteien die Stimme der Wahrheit zu unterscheiden.*

[362] Schnitzler 1967, aphorism 2, p. 18. Source text: *Das Martyrium war immer nur ein Beweis für die Intensität, niemals für die Richtigkeit eines Glaubens.*

[363] Schopenhauer 1920c, paraphrased from p. 374. Source text: *Ein gewisser Grad allgemeiner Unwissenheit ist die Bedienung aller Religionen, ist das Element, in welchem allein sie leben können.*

[364] Schopenhauer 1920c, paraphrased from p. 375. Source text: *Glauben und Wissen sich verhalten wie zwei Schaalen [sic] einer Waage: in dem Masse [sic], als die eine steigt, sinkt die andere.*

[365] Schopenhauer 1920c, paraphrased from p. 361. Source text: *Die Religion mit dem unbefangenen, edlen Streben nach reiner Wahrheit stets in Konflikt gerathen [sic] ist und es immer von Neuem wird.*

366 Seneca the Younger 1985, pp. 152-153. Source text: *Ratio id iudicare vult quod aequum est; ira id aequum videri vult quot iudicavit.*

367 Sextus Empiricus 1933, paraphrased from pp. 10-11. Source text: *ἀλλὰ μὴ δογματίζειν λέγομεν καθ' ὃ δόγμα εἶναί φασί τινες τήν τινι πράγματι τῶν κατὰ τὰς ἐπιστήμας ζητουμένων ἀδήλων συγκατάθεσιν.*

368 Shakespeare 2003a, 2.2.45, p. 153.

369 Shaw 1958, p. 183.

370 Shaw 1921, p. lxxxix.

371 Shaw 1957, pp. 99-100. Source text: *The fact that a believer is happier than a skeptic is no more to the point than the fact that a drunken man is happier than a sober one.*

372 Shelley 1904, p. 196.

373 Shermer 2019, not paginated.

374 Goodstein 2010, not paginated.

375 Sim 2006, p. 162.

376 Simpson 1964, p. 37.

377 Horne 1917, p. 90.

378 Singer 1981, p. 88.

379 Smith 2003, p. 390.

380 Smith 1991, p. 69.

381 Smith 1976, p. 277.

382 Sowell 1996, p. 4.

383 Spencer 1907, p. 35.

384 Spinoza 1862, p. 240.

385 Spinoza 1862, paraphrased from p. 240.

386 Spinoza 1862, paraphrased from p. 258.

387 Roddenberry 1967a, timestamp 00:24:35.

388 Stanislaw 1962, p. 42.

389 Stanislaw 1968, p. 6.

390 Steinmetz 1922, p. 1089.

391 Stendahl 1854, p. 193. Source text: *Cette religion ôte le courage de penser aux choses inaccoutumées, et défend surtout l'examen personnel, comme le plus énorme des péchés.*

392 Stenger 2011, not paginated.

393 Stenger 2012, p. 322.

394 Stenonis 1910, p. 254. Source text: *Pulchra funt, qvæ videntur, pulchriora, qvæ sciuntur, longe pulcherrima, qvæ ignorantur.*

395 Stewart 2008, not paginated.

[396] Sutcliffe 2008, not paginated.

[397] Tegmark 2018, timestamp 00:42:05.

[398] Tennessee 1925, paraphrased from court transcript p. 313.

[399] Tertullian 1998, pp. 38-39. Source text: Cum odio sui coepit veritas.

[400] Harbottle 1897, p. 455. Source text: *Oὔ τι τὰ πολλὰ γ' ἔπη φρονίμην ἀπεφήνατο δόξαν.*

[401] Aristotle 2003c, constructed quotation from Bekker 983b5-983b30, pp. 501-502.

[402] Thiry 1771, p. 27. Source text: *Si l'ignorance de la nature donna la naissance aux Dieux, la connaissance de la nature est faite pour les détruire.*

[403] Thiry 1772, p. ii. Source text: *Théologie n'est que l'ignorance des causes naturelles réduite en système.*

[404] Thiry 1772, aphorism 2, p. 4. Source text: *Théologie est une insulte continuelle à la raison humaine.*

[405] Tolstoy 1985, p. 90.

[406] Garrison 1929, p. 33.

[407] Turing 1950, p. 443.

[408] Twain 1940, p. 324.

[409] Twain 1989, p. 132.

[410] Tyson 2011, not paginated.

[411] Tyson 2018, not paginated.

[412] Tyson 2001, p. 83.

[413] Tyson 2019, not paginated.

[414] Ubbelohde 1963, p. 132.

[415] Valèry 1943, p. 86. Source text: *Le martyr: j'aime mieux mourir que réflechir.*

[416] Verne 1864, p. 209. Source text: *La science, mon garçon, est faite d'erreurs, mais d'erreurs qu'il est bon de commettre, car elles mènent peu à peu à la vérité.*

[417] Vinci 1883, number 1192, p. 297. Source text: *Chi poco pēsa, molto erra.*

[418] McCurdy 1923, p. 54.

[419] Voltaire 1875, p. 249. Source text: *Brûler un livre de raisonnement, c'est dire, nous n'avons pas assez d'esprit pour lui répondre.*

[420] Voltaire 1900, pp. 2-5. Source text: *Remarquez bien que les nez ont été faits pour porter des lunettes, aussi avons-nous des lunettes. Les jambes sont visiblement instituées pour êtres chassées.*

[421] Voltaire 1900, p. 170. Source text: *Traviallons sans raisonner, dit Martin;*

c'est le seul moyen de rendre la vie supportable.

[422] Wallace 1863, paraphrased from p. 307.

[423] Watterson 1992, p. 29.

[424] Wells 2000, p. 15.

[425] Wesley 1826, p. 245.

[426] Wheeler 1989, paraphrased from section 19.6.

[427] Whewell 1847, p. 62.

[428] Whewell 1847, p. 560.

[429] Whitehead 1956, p. 276.

[430] Whitman 1882, p. 302.

[431] Wilde 1905, p. 188.

[432] Wilkins 1708, p. 67.

[433] Wootton 1950, p. 103.

[434] Wright 1958, p. 8. Source text: *Science asks no questions about the ontological pedigree or a priori character of a theory, but is content to judge it by its performance; and it is thus that a knowledge of nature, having all the certainty which the senses are competent to inspire, has been attained - a knowledge which maintains a strict neutrality toward all philosophical systems and concerns itself not at all with the genesis or a priori grounds of ideas.*

[435] Youngman 1918, p. 112.

[436] Addison 1898, number 243, 8 December 1711, p. 359.

[437] Adler 1979, p. 277.

[438] Adorno 1970, article 18, p. 42. Source text: *Es gibt kein richtiges Leben im falschen.*

[439] Adorno 1970, article 34, p. 67. Source text: *Die fast unlösbare Aufgabe besteht darin, weder von der Macht der anderen, noch von der eigenen Ohnmacht sich dumm machen zu lassen.*

[440] Allen 1990, pp. 413-414.

[441] Allen 1969, p. 26.

[442] Amiel 1893, entry from 6 October 1866, p. 123.

[443] Amorth 2002, not paginated.

[444] Angelou 1980, p. 190.

[445] Harper 1898, p. 853.

[446] Arabi 2017, paraphrased from social media post.

[447] Miller 2006, not paginated.

[448] Aristotle 2003b, Bekker 1314b35-1315a05, p. 517.

[449] Arnold 1903a, p. xiv.

[450] Arnold 1888, p. 330.

[451] Asimov 2004, p. 71.

[452] George 2009, not paginated.

[453] Attenborough 1979, p. 308.

[454] Atwood 1990, paraphrased from p. 13.

[455] Aurelius 2003, p. 144.

[456] Ayer 1968, p. 4.

[457] Bacon 1908, p. 265.

[458] Bacon 1902, aphorism 46, p. 23.

[459] Bacon 1657, p. 277. Source text: *Bookes [sic] must follow sciences, and not sciences, bookes.*

[460] Baker 1988, not paginated.

[461] Bakunin 1895, p. 17.

[462] Baldwin 1962, not paginated.

[463] Jones 2009, p. 168. Source text: *Only God has been replaced, as he has all over the West, with respectability and air-conditioning.*

[464] Barker 2006, p. 266.

[465] Barnes 1933, lecture 1, p. 11.

[466] Barry 1998, number 19, p. 185.

[467] Baruch 1957, p. 320.

[468] Baskerville 1907, p. 117.

[469] Bastiat 1854, p. 128. Source text: *Lorsque la spoliation est devenue le moyen d'existence d'une agglomération d'hommes unis entre eux par le lien social, ils se font bientôt une loi qui la sanctionne, une morale qui la glorifie.*

[470] Bates 1945, pp. 164-165. Likely not a direct quote, but a synthesis by Bates of Bossuet's beliefs as expressed in the collection *Œuvres Oratoires de Bossuet* (1926).

[471] Bayle 1826, paraphrased from p. 138.

[472] Bayle 1911, p. 271. Source text: *C'est une illusion toute pure, que de prétendre qu'un sentiment qui passe de siècle en siècle, et de generation en generation, ne peux être entièrement faux.*

[473] Bayle 1820, p. 247. Source text: *De toutes ces choses, on peut recueillir qu'il est très facile de tromper l'homme en mateière de religion, et très dificile de l'y detromper.*

[474] Beit-Hallahmi 1997, p. 242.

[475] Benson 2005, p. 934.

[476] Bentham 1834, p. 74.

[477] Bentham 1969, p. 45.

[478] Widely attributed to Lavrentiy Beria, but without source.

[479] Berkeley 1747, p. 174.

[480] Bharara 2019, p. 58.

[481] King James Bible Online, Matthew 7:12.

[482] Bierce 1993, not paginated.

[483] Bierce 1993, not paginated.

[484] Bierce 1993, not paginated.

[485] Bierce 1993, not paginated.

[486] Stanton 1887a, p. 91.

[487] Boltzmann 1974, p. 193.

[488] Bonner 1919, p. 73.

[489] Boorstin 1962, p. 183.

[490] Boosler 1987, not paginated.

[491] Boyer 2001, p. 47.

[492] Bradlaugh 1870, p. 21.

[493] Bradlaugh 1889, p. 9.

[494] Smith 1973, not paginated.

[495] Browne 1907, p. 35. Source text: *The religion of one seems madness unto another.*

[496] Buck 1938, p. 226.

[497] Buckle 1929, p. 129.

[498] Bruñel 1983, p. 167. Source text: *Dios y la patria están allí codo con codo. No nos traían más que represión y sangre.*

[499] Burbank 1926, p. 9.

[500] Burns 1896, p. 150.

[501] Butler 1921, pp. 212-213.

[502] Butler 1986, p. 137.

[503] Cabell 1917, p. 72.

[504] Dell 2000, p. 54.

[505] Caesarius of Heisterbach 1591, p. 383. Source text: *Caedite eos. Novit enim Dominus qui sunt eius.*

[506] Calvin 1845a, p. 450.

[507] Carlin 2004, p. 17.

[508] Carter 1985, p. 22.

[509] Origen of Alexandria 1872, p. 125.

[510] Chesnut 2006, p. 304.

[511] Chisholm 1971, p. 181.

[512] Clark 1980, p. 184.

[513] Cohen 1933, p. 90.

[514] Coleridge 1836, p. 232.
[515] Colton 1825, reflection 17, p. 22.
[516] Colton 1825, reflection 172, p. 95.
[517] Colton 1825, reflection 141, pp. 106-107.
[518] Soothill 1910, pp. 261-263.
[519] Conway 1884, p. 168.
[520] Cousins 1953, p. 318.
[521] Peter 1977, p. 121.
[522] Coward 1979, p. 17.
[523] Crisp 2000, p. 312.
[524] Salt 1921, p. 7.
[525] Darwin 2003, p. 304.
[526] Davis 1881, paraphrased from p. 192.
[527] Delay 1999, not paginated.
[528] Darrow 1929, p. 33.
[529] Deloria 1969, p. 105.
[530] Dennett 1995, p. 338. Source text: *People ache to believe that we human beings are vastly different from all other species - and they are right!*
[531] Deschner 1996, p. 29. Source text: *Theologe - einziger Experte ohne Ahnung von seinem Forschungsobjekt.*
[532] Dillahunty 2008, timestamp 01:20:15.
[533] Dillard 1998, p. 123.
[534] Diodorus of Sicily 1933, pp. 8-9. Source text: *εἰ γὰρ ἡ τῶν ἐν ᾅδου μυθολογία τὴν ὑπόθεσιν πεπλασμένην ἔχουσα πολλὰ συμβάλλεται τοῖς ἀνθρώποις πρὸς εὐσέβειαν καὶ δικαιοσύνην.*
[535] Dobzhansky 1945, p. 75.
[536] Doren 1926, p. 205.
[537] Douglass 1847, p. 119.
[538] Douglass 1893, paraphrased from p. 529.
[539] Dylan 1986, not paginated.
[540] Eddington 1929, p. 68.
[541] Einstein 1954, not paginated. Source text: *Die Bibel [ist] eine Sammlung ehrwürdiger aber doch reichlich primitiven Legenden.*
[542] Einstein 1922, not paginated. Source text: *Hier draussen weiss niemand wer ich bin, und ich gelte als vermisst.*
[543] Einstein 1982, pp. 51-52.
[544] Einstein 1982, p. 39.
[545] Einstein 1950, p. 115.

[546] Eliot 1883, p. 117.

[547] Ayer 1968, p. 56.

[548] Lactantius 1543, p. 520. Source text: *Si vult et non potest, imbecillis est. Si potest et non vult, inuidus.*

[549] Euripides 2008, fragment 7, p. 9. Source text: *κρεῖσσον δὲ πλούτου καὶ βαθυσπόρου χθονὸς ἀνδρῶν δικαίων κἀγαθῶν ὁμιλίαι.*

[550] Everson 1947a, p. 330 US 9.

[551] Fallaci 1975, p. 15.

[552] The Nation's Best Bible College Gets Low Grades on Racial Diversity 2001, p. 43.

[553] Festinger 2014, not paginated.

[554] Festinger 2014, not paginated.

[555] Feuerbach 1848, p. 401. Source text: *Das Geheimnis der Theologie [ist] die Anthropologie.*

[556] Feuerbach 1848, p. 402. Source text: Homo homini Deus est.

[557] Feuerbach 1848, p. 407. Source text: *Wenn die Moral auf die Theologie, das Recht auf göttliche Einsetzung gegründet wird, dann kann man die unmoralischsten, unrechtlichsten, schändlichen Dinge rechtfertigen und begründen.*

[558] Flammarion 1923, p. 24.

[559] Foote 1893, p. 190.

[560] Shlachter 1985, p. 7.

[561] France 1926, p. 8. Source text: *Le christianisme a beaucoup fait pour l'amour en en faisant un péché.*

[562] France 1906, p. 18. Source text: *S'il fallait absolument choisir, j'aimerais mieux faire une chose immorale qu'une chose cruelle.*

[563] Koser 1908b, p. 183. Source text: *Je crois qu'il vaut mieux garder un silence parfait sur ce qui regarde les fables chrétiennes, canonisées par leur ancienneté et par la crédulité des hommes, puériles et absurdes pour des hommes qui raisonnent conséquemment.*

[564] Koser 1908a, p. 13. Source text: *A l'égard des théologiens, il parait qu'ils se ressemblent tous en général, de quelque religion ou de quelque nation qu'ils soient; leur dessein est toujours de s'arroger une autorité despotique sur les consciences. Cela suffit pour les rendre persécuteurs zélés de tous ceux dont la noble hardiesse ose dévoiler la vérité.*

[565] Freud 1927, p. 71. Source text: *Bringt sie einerseits Zwangseinschränkungen, wie nur eine individuelle*

Zwangsneurose, so enthält sie anderseits ein System von Wunschillusionen mit Verleugnung der Wirklichkeit, wie wir es isoliert nur bei einer Amentia, einer glückseligen halluzinatorischen Verworrenheit, finden.

566 Freud 1930, p. 120. Source text: *Die Religionen wenigstens haben die Rolle des Schuldgefühls in der Kultur nie verkannt. Sie treten ja, was ich anderer Stelle nicht gewürdigt hatte, auch mit dem Anspruch auf, die Menschheit von diesem Schuldgefühl, das sie Sünde heissen, zu erlösen.*

567 Freud 1921, p. 55. Source text: *Darum muss eine Religion, auch wenn sie sich die Religion der Liebe heisst, hard und lieblos gegen diejenigen sein, die ihr nicht angehören.*

568 Fry 2004, p. 61.

569 Fry 2004, p. 59.

570 Gage 1893, p. 539.

571 Gage 1893, p. 228.

572 Fahie 1903, paraphrased from pp. 319-321.

573 Gandhi 1927, p. 1285.

574 Garrison 1885, p. 188.

575 Gawande 2016, paraphrased from online article.

576 Debate on Modern Abolitionism 1836, p. 5.

577 Gervais 2013, not paginated.

578 Vahdat 2019, not paginated.

579 Gibbon 2003, paraphrased from p. 12. Also attributed to Seneca the Younger.

580 Ferraù 2012, not paginated. Source text: *Una politica concepita solamente come conquista e mantenimento del potere è la cosa più ostile al desiderio di bene dell'uomo e al suo rapporto con l'infinito.*

581 Goethe 1944, p. 90. Source text: *Die Kirch [sic] allein, meine lieben Frauen, kann ungerechtes Gut verdauen.*

582 Goldman 1913, p. 42.

583 Goodall 2002.

584 Gould 1997, p. 216.

585 Grant 1870, not paginated.

586 Bragg, The Iliad, 13 September 2018, timestamp 00:13:30.

587 Green 2017, not paginated.

588 Haider 2019, not paginated.

589 Dilliard 1952, p. 153.

[590] Harris 2005b, timestamp 00:17:50.
[591] Harris 2011, timestamp 01:06:57.
[592] Pinto 2017, not paginated.
[593] Hayek 2011, p. 213.
[594] Hayes 1923, p. 144.
[595] Heinlein 1974, p. 244.
[596] Hellman 1941, p. 8.
[597] Herbert 1976, p. 268.
[598] Herbert 1841, p. 145.
[599] Herman 1991, paraphrased from p. xiii. Source text: *One of the assumptions of this book is that religions and philosophies exist primarily to solve human problems.*
[600] Hirsi Ali 2010, pp. 212-213.
[601] Hirsi Ali 2006, not paginated.
[602] Hitchens 2007, p. 266.
[603] Hobsbawn 1994, paraphrased from timestamp 00:10:55 to 00:13:45.
[604] Hoffer 1974, reflection 13, p. 9.
[605] Hoffer 1974, reflection 13, p. 9.
[606] Hoffer 1963, p. 147.
[607] Hofstadter 1985, p. 756.
[608] Holmes 1919, p. 40
[609] Holmes 1919, p. 41
[610] Hook 1964, p. 141.
[611] Hook 1964, p. 141.
[612] Hubbard 1907, p. 88.
[613] Hugo 1878, p. 63. Source text: *Avant la Révolution, messieurs, la construction sociale était ceci: En bas le peuple; au-dessus du peuple, la religion, représentée par le clergé; a côté de la religion, la justice, représentée par la magistrature. Et, à ce moment de la société humaine, qu'était-ce que le peuple? C'était l'ignorance. Qu'était-ce que la religion ? C'était l'intolérance. Et qu'était-ce que la justice? C'était l'injustice.*
[614] Hume 1767, paraphrased from p. 454. Source text: *The heroes of paganism correspond exactly to the saints in popery and holy dervishes in Mahometanism. The place of Hercules, Theseus, Hector, Romulus is now supplied by Dominic, Francis, Anthony, and Benedict.*
[615] Hume 1960, paraphrased from p. 483.
[616] Holy 1996, p. 40.

[617] Huxley 1957, p. 220.

[618] Huxley 1961, p. 14.

[619] Huxley 1888, p. 312.

[620] Ibsen 1999, p. 59.

[621] Ingersoll 1909a, p. 175.

[622] Ingersoll 1909c, p. 518.

[623] Ingersoll 1909b, title page.

[624] Ingersoll 1902b, p. 257.

[625] Woods 2019, not paginated.

[626] Human Rights Lawyer Nasrin Sotoudeh 2009, not paginated.

[627] Jacobs 2004, p. 115.

[628] James 1902, p. 491.

[629] Je suis Charlie 2015, front page.

[630] Jefferson 1832, p. 168.

[631] Jefferson 1903, query 17, p. 222.

[632] Jefferson 1903, query 17, p. 221.

[633] Jefferson 1816b, p. 1.

[634] Kaku 2010, not paginated.

[635] Taber 1897, p. 86.

[636] Kant 1870, p. 20. Source text: *Ich soll niemals anders verfahren, als so, dass ich auch wollen könne, meine Maxime solle ein allgemeines Gesetz werden*.

[637] King 1965, timestamp 00:02:35.

[638] Knight 1970, p. 156.

[639] Knight 1955, p. 29.

[640] Lateran Treaty 1929, article 8. Source text: *Le offese e le ingiurie pubbliche commesse nel territorio italiano contro la persona del Sommo Pontefice con discorsi, con fatti e con scritti sono punite come le offese e le ingiurie alla persona del Re*.

[641] Latham 1878, p. 398.

[642] Lazarus 1888, p. 203. Written in 1883 for the *Bartholdi Pedestal Fund for the Statue of Liberty*.

[643] Lecky 1886, p. 40.

[644] Lecky 1886, p. 40.

[645] Lennon 1971, not paginated.

[646] Sime 1879, paraphrased from p. 101.

[647] Lewis 1938, not paginated.

[648] Lewis 1938, not paginated.

[649] Lincoln 1864, p. 1.

[650] Lindsey 1935, p. 280.
[651] Lippman 1964, pp. 129-130.
[652] Locke 2003a, p. 26.
[653] Locke 2003b, Epistle to the Reader, p. 85.
[654] Locke 2010, p. 99.
[655] Pfeiffer 1965, not paginated.
[656] Lucretius 1910, p. 29.
[657] Luther 1973, p. 6.
[658] Luther 1967, p. 25. Also King James Bible Online, Romans 13:1 and 1 Peter 2:18.
[659] Macaulay 1860, p. 331.
[660] Maistre 1880, pp. 80-81. Source text: *Le christianisme a été prêché par des ignorants et cru par des savants, et c'est en quoi il ne ressemble à rien de connu.*
[661] Mann 1967. Source text: Prägen Sie sich immmerhin ein, dass Toleranz zum Verbrechen wird, wenn sie dem Bösen gilt.
[662] Martineau 1837, p. 105.
[663] Marx 2014, p. 3. Source text: *Der Mensch macht die Religion, die Religion macht nicht den Menschen.*
[664] Maugham 1915, p. 180.
[665] May 1953, p. 150.
[666] May 1953, paraphrased from p. 192.
[667] McCabe 1930, p. 58. Source text: *I doubt if France ever did a greater service to its women than when it (though not entirely) closed their convents.*
[668] McCreary 2005, not paginated.
[669] McCreary 2005, not paginated.
[670] Bragg, Science and Religion, 29 October 1998, timestamp 00:25:00.
[671] McGoohan 1967, timestamp 00:21:30.
[672] Overstreet 1964, p. 149.
[673] McLuhan 1977, not paginated.
[674] Mencken 1908, p. 96.
[675] Mencken 1949, p. 617.
[676] Mencken 1949, p. 616.
[677] Mencken 1949, p. 624.
[678] Mencken 1956, aphorism 418, p. 282.
[679] Mencken 1916, aphorism 1, p. 73.
[680] Merkel 2009, not paginated. Source text: *Toleranz ist Ausdruck des Respekts vor der Geschichte, der Tradition, der Religion und der*

Identität anderer. Doch niemand sollte sich täuschen: Toleranz bedeutet nicht Beliebigkeit. Null Toleranz muss es für die geben, die die unveräußerlichen Rechte des Menschen missachten und sie mit Füßen treten.

681 Voltaire 1801, paraphrased from p. xvi. Source text: *Mais aussi combien n'ai-je pas souffert en moi-même, lorsque j'ai été forcé de vous prêcher ces pieux mensonges que je détestais dans le coeur! Quel mépris n'avais-je pas pour mon ministère, et particulièrement pour cette superstitieuse messe, et ces ridicules administrations de sacrements, surtout lorsqu'il fallait les faire avec cette solennité qui attirait votre piété et toute votre bonne foi! Que de remords ne m'a point excités votre crédulité! Mille fois sur le point d'éclater publiquement, j'allais dessiller vos yeux; mais une crainte supérieure à mes forces me contenait soudain, et m'a forcé au silence jusqu'à ma mort.*

682 Voltaire 1801, p. 8. Source text: *Saint Jerome dit formellement en plusieurs endroits de ses prologues, qu'elles ont été corrompues et falsifiées, étant déjà de son temps entre les mains de toutes sortes de personnes, qui y ajoutaient et retranchaient tout ce que bon leur semblait, ensorte qu'il y avait, dit-il, autant d'exemplaires différens qu'il y avait de différents copies.*

683 Mettrie 1921, p. 111. Source text: *Si l'athéisme, disait-il, était généralement répandu, toutes les branches de la religion seraient alors détruites et coupées par la racine. Plus de guerres théologique, plus de soldats de religion, soldats terribles.*

684 Roddenberry 1967b, timestamp 00:47:12.

685 Mill 2003, p. 291.

686 Mill 1867, p. 74.

687 Mills 1958, p. 200.

688 Mises 1947, p. 30.

689 Mitchell 1979, not paginated.

690 Mokyr 2016, p. 132.

691 Montaigne 2003, p. 149.

692 Montesquieu 1869, letter 27, p. 58. Source text: *Aussi puis-je t'assurer qu'il n'y a jamais eu de royaume où il y ait eu tant de guerres civiles que dans celui de Christ.*

693 Montesquieu 1869, letter 33, p. 67. Source text: *Crois-tu qu'ils soient condamnés à des châtiments éternels, et que Dieu les punisse pour n'avoir pas pratiqué une religion qu'il ne leur a pas fait*

connaître?

[694] Montesquieu 1869, letter 83, p. 172. Source text: *J'avoue que les histoires sont remplies des guerres de religion; mais qu'on y prenne bien garde, ce n'est point la multiplicité des religions qui a produit ces guerres, c'est l'esprit d'intolérance qui animait celle qui se croyait la dominante.*

[695] Elliot 1861, p. 393.

[696] Murray 1940, p. 179.

[697] Niebuhr 1953, p. 109.

[698] Nietzsche 1917, aphorism 483, p. 363. Source text: *Überzeugungen sind gefährlichere Feinde der Wahrheit als Lügen.*

[699] Nietzsche 1919a, dictum 7, p. 62. Source text: *Ist der Mensch nur ein Fehlgriff des Gottes? Oder ist Gott nur ein Fehlgriff des Menschen?*

[700] Nietzsche 1921, paraphrased from aphorism 14, p. 200. Source text: *Vielleicht bildet sich die Ameise im Walde ebenso stark ein, dass sie Ziel und Absicht für die Existenz des Waldes ist, wie wir diess thun [sic], wenn wir an den Untergang der Menschheit in unserer Phantasie fast unwillkürlich den Erduntergang anknüpfen.*

[701] Nietzsche 1921, aphorism 22, p. 22. Source text: *Im Anfang war der Unsinn, und der Unsinn war, bei Gott!, und Gott (göttlich) war der Unsinn.*

[702] Nine Inch Nails 1994, not paginated.

[703] Nisbet 1982, pp. 182-183.

[704] North Carolina 1930, not paginated.

[705] Norton 1913, p. 248.

[706] O'Connor 1983, p. 291.

[707] Gregory 2007, as attributed to Leslie Orgel by Francis Crick, not paginated.

[708] Ouida 1895, p. 129.

[709] Paine 1896, p. 23.

[710] Palmer 1819, pp. 8-9.

[711] Drennan 1995, p. 120.

[712] Holmes 1919, p. 18.

[713] Parks 2005, not paginated.

[714] Parks 1992, p. 30.

[715] Pascal 1854, p. 358. Source text: *Tout ce qu'il y a d'hommes sont presque toujours emportés à croire non pas par la preuve, mais par l'agrément.*

[716] Pascal 2003, pensée 895, p. 347.

[717] Pascal 2003, pensée 489, p. 259.

[718] Passmore 1970, p. 289.

[719] Pérez-Reverte 2006, p. 160.

[720] Pike 1964, p. 39.

[721] Plato 2003b, Stephanus 409, p. 337.

[722] Pope John Paul II 1996, paraphrased from article 5. Source text: *Las teorías de la evolución ... son incapaces de fundar la dignidad de la persona.*

[723] Ayer 1968, p. 292.

[724] Popper 1947, p. 226.

[725] Popper 1947, p. 226.

[726] Pratchett 2003, p. 192.

[727] Plato 2003c, Stephanus 160, p. 522; Stephanus 168, p. 526; Stephanus 170, p. 527; Stephanus 171, p. 528; Stephanus 183, p. 534.

[728] Stanton 1887a, p. 804.

[729] Pullman 2002, not paginated.

[730] The Noble Qur'an, 109:6.

[731] Jacobsen 2017, not paginated.

[732] Ra 2015a, not paginated.

[733] Brown 1963, p. 290.

[734] Rand 1990, p. 184.

[735] Rand 1967, p. 17.

[736] Rawls 1971, p. 134.

[737] Rawls 1971, p. 217.

[738] Reeves 2019, not paginated.

[739] Rich 1975, p. 12.

[740] Roberts, Stephen. [www.askatheists.com/user/5972]

[741] Schwarz 1992, not paginated.

[742] Rochefoucauld 1923, maxim 347, p. 166. Source text: *Nous ne trouvons guère de gens de bon sens, que ceux qui sont de notre avis.*

[743] Guillen 2004, p. 90.

[744] Alexander 1991, not paginated.

[745] Peter 1977, p. 67.

[746] Rousseau 2003, p. 438.

[747] Rousseau 2003, p. 439.

[748] Rousseau 2003, p. 397.

[749] Ruse 1982, p. 58.
[750] Rushdie 1990, p. 3.
[751] Rushdie 1990, p. 3.
[752] Rushdie 1989, p. 95.
[753] Ruskin 1890, p. 195.
[754] Russell 1930, p. 50. Source text: *The fact that an opinion has been widely held is no evidence whatever that it is not entirely absurd; indeed, in view of the silliness of the majority of mankind, a widespread belief is more likely to be foolish than sensible.*
[755] Russell 1950, p. 104.
[756] Russell 1957, p. 21.
[757] Russell 1959, p. 108.
[758] Russell 1945, p. 492.
[759] Russell 1958, p. 162.
[760] Sagan 1979, pp. 332-333.
[761] Sagan 1996, not paginated.
[762] Sagan 1994a, paraphrased from pp. 6-7.
[763] Sagan 1987, not paginated.
[764] Sagan 1996, not paginated.
[765] Sakharov 1968, p. 29.
[766] Salt 1921, p. 213.
[767] Sand 1873, p. 132. Source text: *J'aime mieux croire que Dieu n'existe pas que de le croire indifférent.*
[768] Sanger 1921, p. 172.
[769] Sanger 1914, p. 1.
[770] Santayana 1931a, essay 23, p. 48.
[771] Santayana 1931a, essay 23, p. 48.
[772] Santayana 1931b, essay 107, p. 272.
[773] Sarton 1988, p. 47.
[774] Sartre 1965, p. 19.
[775] Schenck 1919, p. 249 US 52.
[776] Schick 2012, not paginated. Source text: *Wer die Seele der Gläubigen mit Spott und Hohn verletzt, der muss in die Schranken gewiesen und gegebenenfalls auch bestraft werden.*
[777] Schlegel 1904, paraphrased from p. 126. Source text: *Es gibt Menschen, deren ganze Tätigkeit darin besteht, immer Nein zu sagen. Es wäre nichts kleines immer recht Nein sagen zu können, aber wer weiter nichts kann, kann er gewiss nicht recht. Der Geschmack dieser Neganten ist eine tüchtige Schere, um die*

Extremitäten des Genies zu säubern; ihre Aufklärung eine grosse Lichtputze für die Flamme des Enthusiasmus; und ihre Vernunft ein gelindes Laxative gegen unmässige Lust und Liebe.

778 Schlesinger 1989, not paginated.

779 Schlesinger 1989, not paginated.

780 Schopenhauer 1920b, paraphrased from p. 395-396. Source text: *Die Absurditäten im Dogma sind eben das Stempel und Abzeichen des Allegorischen und Mythischen.*

781 Schopenhauer 1920a, paraphrased from p. 1285. Source text: *Alle Religionen und Philosophien allein den Tugenden des Willens, oder Herzens, einen Lohn in der Ewigkeit zuerkennen, nicht denen des Intellekts, oder Kopfes.*

782 Seaver 1949, p. 342.

783 Sen 2006, pp. 9-10.

784 Seneca the Younger 2002, pp. 388-389. Source text: *Cui prodest scelus, is fecit.*

785 Servetus 1969, p 264.

786 Shakespeare 2003b, 1.2.140, p. 570.

787 Kanso 2018, not paginated.

788 Shaw 1919, p. 262.

789 Shaw 1957, p. 56.

790 Sheed 1985, p. 92.

791 Shelley 1915b, p. 45.

792 King James Bible Online, Matthew 10:34.

793 The Instructions of Shuruppag 2001, not paginated.

794 Sinclair 1918, p. 23.

795 Singer 1981, p. 3.

796 Bragg, Evolution, 15 April 1999, timestamp 00:17:35.

797 Smith 1994, p. 101.

798 Smith 1957, p. 296.

799 Spencer 1898, paraphrased from p. 48.

800 Spinoza 2003, paraphrased from p. 663.

801 Stanislaw 1962, p. 66.

802 Stanislaw 1968, p. 9.

803 Stanton 1887b, p. 124.

804 Stanton 1993, p. 7.

805 Stendahl 2016, p. 164. Source text: *J'ai assez vécu pour voir que la différence engendre la haine.*

806 Storr 1991, pp. 156-157.

[807] Stowe 1896, p. 374.

[808] Suarès 1915, p. 33. Source text: *L'hérésie est la vie de la religion. C'est la foi qui fait les hérétiques. Dans une religion morte, il n'y a plus d'hérésies.*

[809] Ballou 1881, p. 433.

[810] Swift 1982, p. 85.

[811] Syrus 1954, pp. 62-63. Source text: *Male facere qui vult numquam non causam invenit.*

[812] Thatcher 1975, not paginated.

[813] Thatcher 1965, not paginated.

[814] Thiry 1771, pp. 217-218. Source text: *Quand nous voudrons remonter à l'origine des choses, nous trouverons toujours que c'est l'ignorance et la crainte qui ont créé les Dieux; que c'est l'imagination, l'enthousiasme et l'imposture qui les ont ornés ou défigurés; que c'est la faiblesse qui les adore; que c'est la crédulité qui les nourrit; que c'est l'habitude qui les respecte; que c'est la tyrannie qui les soutient; afin de profiter de l'aveuglement des hommes.*

[815] Thiry 1772, aphorism 119, pp. 151-152. Source text: *On croit fermer la bouche à ceux qui nient l'existence d'un Dieu, en leur disant que tous les hommes, dans tous les siècles, dans tous les pays ont reconnu l'empire d'une Divinité quelconque: qu'il n'est point de peuple sur la terre qui n'ait eu la croyance d'un être invisible et puissant, dont il a fait l'objet de son culte et de sa vénération: enfin qu'il n'est pas de nation, si sauvage qu'on la suppose qui ne foit persuadée de l'existence de quelque intelligence supérieure à la nature humaine. Mais la croyance de tous les hommes peut-elle changer une erreur de vérité?*

[816] Thucydides 2003, p. 354.

[817] Reichskonkordat 1933, article 16. Source text: *Vor Gott und auf die heiligen Evangelien schwöre und verspreche ich, so wie es einem Bischof geziemt, dem Deutschen Reich und dem Lande, Treue. Ich schwöre und verspreche, die verfassungsmäßig gebildete Regierung zu achten und vorn meinem Klerus achten zu lassen. In der pflichtmäßigen Sorge (um das Wohl und das Interesse des deutschen Staatswesens) werde ich in Ausübung des mir übertragenen geistlichen Amtes jeden Schaden zu verhüten trachten, der es bedrohen könnte.*

[818] Tuchman 2014, not paginated.

[819] Tuchman 1979, p. 327.

[820] Twain 1974, p. 24.

[821] Universal Declaration of Human Rights 1948, article 18.

[822] Universal Declaration of Human Rights 1948, article 19.

[823] Voltaire 1816b, p. 147. Source text: *Toute secte, en quelque genre que ce puisse être, est le ralliement du doute et de l'erreur.*

[824] Voltaire 1874, epistle 49, p. 194. Source text: *Si Dieu n'existait pas, il faudrait l'inventer.*

[825] Voltaire 1816a, paraphrased from p. 94.

[826] Vonnegut 1999, p. 9.

[827] Miller 2004, not paginated.

[828] Washington 1906, p. 165.

[829] Watson 1796, p. 32.

[830] King James Bible Online, Numbers 7:7-10.

[831] Waugh 1962, p. 86.

[832] Weil 1981, p. 11.

[833] Weinberg 2006, timestamp 00:28:01.

[834] Rittner 2006, page x.

[835] Wiesel 1992, p. 5.

[836] Westermarck 1917, p. 746.

[837] Whitehead 1927, p. 26.

[838] Wilde 1905, p. 55.

[839] Wilde 1899, p. 112.

[840] McCormack 1997, p. 101.

[841] Williams 2002, p. 52.

[842] Wilson 1968, not paginated.

[843] An Intellectual Entente 2009, not paginated.

[844] Wollstonecraft 1833, p. 76.

[845] Wright 1998, p. 356.

[846] Wu 1963, p. 95.

[847] Wu 1963, p. 95.

[848] The Zend-Avesta, fargard 18.2.17(41), p. 193.

[849] Zinn 1999, p. 11.

[850] Zinn 1990, p. 105.

[851] Abrams 1919, p. 250 US 630.

[852] Acton 1919a, p. 504.

[853] Acton 1919b, p. 4.

[854] Acton 1919b, p. 97.

[855] Adams 1780, Article 16.

[856] Adams 1788, p. 475.

[857] Bernard 1320, not paginated.

[858] Archer 1907, p. 80.

[859] Arendt 1962, p. 350.

[860] Arendt 1962, p. 382.

[861] Arendt 1962, p. 474. Source text: *The ideal subject of totalitarian rule is not the convinced Nazi or the convinced Communist, but people for whom the distinction between fact and fiction (i.e., the reality of experience), and the distinction between true and false (i.e., the standards of thought), no longer exists.*

[862] Aristotle 2003b, Bekker 1291b30, p. 491.

[863] Aristotle 2003b, Bekker 1295b35, p. 496.

[864] Aristotle 2003b, Bekker 1295b6, p. 495.

[865] Asimov 1980, paraphrased from p. 19.

[866] Badawi 2015.

[867] Bakunin 1895, p. 15.

[868] Baldwin 2011, not paginated.

[869] Baldwin 2007, p. 149.

[870] Balzac 1892, p. 403. Source text: *La liberté politique, la tranquillité d'une nation, la science même, sont des présents pour lesquels le destin prélève des impôts de sang.*

[871] Malala Yousafzai Rallies 2013, not paginated.

[872] Barnhill 1914, p. 34.

[873] Bayle 1826, paraphrased from p. 138.

[874] Beccaria 1775, p. 167.

[875] Bellamy 1913, p. 26.

[876] Berlin 2002, p. 345.

[877] Bernanos 1982, not paginated. Source text: *Les démocraties ne peuvent pas plus se passer d'être hypocrites que les dictatures d'êtres cyniques.*

[878] Bernanos 1968, p. 117.

[879] Bhutto 2007, not paginated.

[880] Bierce 1993, not paginated.

[881] Blackstone 1768, p. 269.

[882] Copeland 1999a, paraphrased from p. 386.

[883] Bonner 1919, p. 73.

[884] Bossuet 1821, article 2, p. 66. Source text: *L'autorité royale est sacrée.*

[885] Bossuet 1821, article 1, p. 96. Source text: *L'autorité royale est absolue.*

[886] Bryan 1909, p. xxxv.

[887] Buckley 2004, not paginated.

[888] Hart 1966, p. 171.

[889] Bulmer-Thomas 1949, p. 241.

[890] Burke 1782, p. 63.

[891] Gould 1905, p. 26.

[892] Bury 1913, p. 251.

[893] Calhoun 1848, paraphrased from p. 52, column 1. Source text: *But of the few nations who have been so fortunate as to adopt wise constitutions, still fewer have had the wisdom long to preserve them. It is harder to preserve them than to obtain liberty.*

[894] Camus 1947, p. 149. Source text: *Mais il vient toujours une heure dans l'histoire où celui qui ose dire que deux et deux font quatre est puni de mort.*

[895] Andrews 1989, p. 67.

[896] Chesnut 2006, p. 9.

[897] Chomsky 2003, p. 43.

[898] Eade 1945, p. 275.

[899] Churchill 1947, not paginated.

[900] Cicero 2006, p. 133. Source text: *Si unus quisque nostrum ad se rapiat commoda aliorum detrahatque, quod cuique possit, emolumenti sui gratia, societas hominum et communitas evertatur necesse est.*

[901] Cerf 1945, p. 258.

[902] Cleyre 2004, p. 102.

[903] Collins 1713, p. 36.

[904] Colman 1891, p. 6.

[905] Colman 1891, p. 7.

[906] Condorcet 1847, paraphrased from pp. 121-122. Source text: *L'habitude peut familiariser les hommes avec la violation de leurs droits naturels ... Par exemple, tous n'ont-ils pas violé le principe de l'égalité des droits, en privant tranquillement la moitié du genre humain de celui de concourir à la formation des lois, en excluant les femmes du droit de cité? ... Pour que cette exclusion ne fût pas un acte de tyrannie, il faudrait ou prouver que les droits naturels des femmes ne sont pas absolument les mêmes que ceux des hommes, ou montrer qu'elles ne sont pas capables de les exercer.*

[907] Conklin 1937, p. 601.

908 Butt 2009, not paginated.
909 Cromwell 1901, p. 430. Source text: I do dissolve this Parliament.
910 Darrow 1920, p. 39.
911 Darrow 1929, p. 39.
912 Debs 1994, p. 251.
913 Dewey 1910b, p. 2.
914 Diderot 1819, p. 10.
915 Diderot 1821a, p. 5. Source text: *Du fanatisme à la barbarie, il n'y a qu'un pas.*
916 Disraeli 1879, p. 70.
917 Douglass 1857, p. 22.
918 Durant 2012, not paginated.
919 Einstein 1987, p. 310. Source text: *Autoritätsdusel ist der grösste Feind der Wahrheit.*
920 Emerson 1910, journal entry from 20 June 1831, p. 386.
921 Epictetus 2003, p. 131.
922 Emmerton 1899, p. 377.
923 Everson 1947a, p. 330 US 15.
924 Everson 1947b, p. 330 US 26.
925 Ferber 1963, pp. 136-137.
926 Forster 1979, p. 70.
927 Jameson 1906, p. 618.
928 Friedman 1980, p. 13.
929 Friedman 1980, paraphrased from p. 148.
930 Fromm 1994, p. 6.
931 Fulbright 1966, pp. 3-4.
932 Furet 1999, pp. 205-206.
933 Gage 1893, p. 540.
934 Gandhi 1922, p. 43.
935 Gandhi 1924, p. 259. Source text: I*n this age of the rule of brute force, it is almost impossible for anyone to believe that any one else could possibly reject the law of the final supremacy of brute force.*
936 Frazier 1998, p. 37.
937 Biskupski 1990, paraphrased from pp. 132, 134, 135. Source text: *A iż w Rzeczypospolitej naszej jest dissidium niemałe in causa religionis christianae, zabiegając temu, aby się z tej przyczyny między ludźmi seditio jaka szkodliwa nie wczęła, którą po inszych królestwach jaśnie widzimy, obiecujemy to sobie spoinie pro nobis et successoribus nostris in perpetuum sub vinculo iuramenti, fide,*

honore et conscientiis nostris, iż którzy jesteśmy dissidentes de religione, pokój między sobą zachować a dla różnej wiary i odmiany w kościelech krwie nie przelewać, ani się penować confiscatione-bonorum, poczciwością, carceribus et exilio, i zwierzchności żadnej, ani urzędowi do takowego progressu żadnym sposobem nie pomagać.

938 Gibbon 2003, p. 24.

939 Green 1977, p. 5.

940 Green 2016, not paginated.

941 Haldane 1963, p. 464.

942 Hamilton 1904, p. 411.

943 Hamilton 2003, number 51, p. 164.

944 Dilliard 1952, p. 190.

945 Hayek 1944, p. 174.

946 Hazlitt 1934, p. 201.

947 Hazlitt 1902a, p. 279.

948 Heinlein 1961, p. 185.

949 Heinlein 1986, pp. 333-334.

950 Hemingway 1995, p. 66.

951 Henry 1860, p. 2.

952 Herodotus 2003, p. 97.

953 Herzen 1981, p. 135.

954 Hobbes 1953, paraphrased from p. xl.

955 Hoffer 1974, reflection 13, p. 9.

956 Hoffer 1963, p. 91.

957 Avni 1989, p. 16.

958 Holmes 1891, p. 190.

959 Gutnick 2017, not paginated.

960 Smith 1918, letter 574, p. 168. Source text: *Besides this, they will provide in all the book-sellers' shops and in the printing presses, so far as this is possible, that in future no new work shall be printed, and especially that no libelous books shall be sold, either publicly or privately; and all the rulers will make a law that if anyone wishes to publish, sell or print any new work, such works must first be revised and reviewed by certain approved and learned men, who have skill in letters, and unless they have the permission and approbation of these men they are not to be sold, printed or published.*

961 Hook 1940, p. 9.

[962] Hook 1940, p. 10.

[963] Horace Mann League 1966, not paginated.

[964] Horkheimer 1947, p. 24.

[965] Hallerberg 1977, p. 137.

[966] Hume 1889, p. 135.

[967] Hume 1912, p. 117.

[968] Huxley 1958, p. 12.

[969] Ibsen 1999, p. 68.

[970] Iles 1918, p. 177.

[971] Ingersoll 1894, paraphrased from pp. 17-18.

[972] Ingersoll 1902a, p. 140.

[973] Jefferson 1787, p. 1.

[974] Jefferson 1789, p. 1.

[975] Jefferson 1814, p. 1.

[976] Jefferson 1816a, p. 4.

[977] Jefferson 1804, p. 1.

[978] Joyce 1989, p. 173.

[979] Hayman 2001, p. 360.

[980] Kant 1920, p. 7. Source text: *Ich gestehe frei: die Erinnerung des David Hume war eben dasjenige, was mir vor vielen Jahren zuerst den dogmatischen Schlummer unterbrach und meinen Untersuchungen im Felde der spekulativen Philosophie eine ganze andere Richtung gab.*

[981] Kennedy 1961, not paginated.

[982] Kennedy 1962b, p. 10.

[983] Kennedy 1960, p. 6.

[984] Copeland 1999b, p. 139.

[985] Khaldun 1958, p. 313. Paraphrased from source text: *Now, royal authority is a noble and enjoyable position. It comprises all the good things of the world, the pleasures of the body, and the joys of the soul. Therefore, there is, as a rule, great competition for it. It rarely is handed over voluntarily, but it may be taken away.*

[986] King 1960, p. 507.

[987] King 1963, p. 5.

[988] Koestler 1943, p. 5.

[989] Laden 2002, not paginated.

[990] Fecteau 1945, p. 92.

[991] Lamartine 1941, p. 797. Source text: *Je suis citoyen de tout homme qui pense. La vérité, c'est mon pays.*

[992] Landor 1891, p. 379.

[993] Lee 1856, not paginated.

[994] Lewis 1935, p. 245.

[995] Lincoln 1898, p. 94.

[996] Lippman 1944, p. 89.

[997] Farr 1985, paraphrased from p. 390.

[998] Locke 2003a, p. 37.

[999] Locke 2003a, p. 71.

[1000] Luxemburg 1922, p. 109. Source text: *Freiheit nur für die Anhänger der Regierung, nur für die Mitglieder einer Partei - mögen sie noch so zahlreich sein - ist keine Freiheit. Freiheit ist immer Freiheit des anders Denkenden.*

[1001] Machiavelli 1891, pp. 188-189. Source text: *Per il che si ha a notare che gli uomini si debbono o vezzeggiare o spegnere, perché si vendicano delle leggieri offese.*

[1002] Madison 1965, p. 9.

[1003] Madison 1965, p. 11.

[1004] Maistre 1853, p. 266. Source text: *Toute nation a le gouvernement qu'elle mérite.*

[1005] Malcolm X 1962, p. 8.

[1006] Mandela 1962, not paginated.

[1007] Mandela 2013, p. 67.

[1008] Mandela 2013, p. 68.

[1009] Mandela 2013, p. 98.

[1010] Mann 1969b, p. 250.

[1011] Mason 1776, section 12.

[1012] Province of Massachusetts 1742, paraphrased from p. 81.

[1013] Commonwealth of Massachusetts 2019, not paginated.

[1014] May 1878, p. 140.

[1015] McCabe 1933, p. 4.

[1016] Mencken 1956, aphorism 330, p. 222.

[1017] Mencken 2006, p. 21.

[1018] Mencken 2006, p. 21.

[1019] Merkel 2009, not paginated.

[1020] Mill 2003, p. 300.

[1021] Mill 2003, p. 274.

[1022] Mill 2003, p. 274.

[1023] Mill 2003, p. 289.

[1024] Milton 1911, p. 12.

[1025] Milton 1911, p. 3.

[1026] Erskine 1792, p. 105.

[1027] Montesquieu 1869, letter 58, p. 118. Source text: *Le zèle pour les progrès de la Religion est différent de l'attachement qu'on doit avoir pour elle, et que pour l'aimer et l'observer il n'est pas nécessaire de haïr et de persécuter ceux qui ne l'observent pas.*

[1028] Montesquieu 2003, paraphrased from p. 70.

[1029] Moore 1861, p. 101.

[1030] Morley 1917, p. 191.

[1031] Morrison 2012, p. 2.

[1032] Murrow 1967, p. 349.

[1033] New York Times 1971, p. 403 US 717.

[1034] New York Times 1971, p. 403 US 717.

[1035] More Than 1000 People 2019, not paginated.

[1036] Niebuhr 1953, p. 96.

[1037] Niebuhr 1945, p. vi.

[1038] Nietzsche 1922a, aphorism 125, p. 232. Paraphrased from source text: *Dabei kann es doch an vielen Orten Europa's ihrerseits zu gelegentlichen Handstreichen und Überfallen kommen: dem nächsten Jahrhundert wird es hie [sic] und da gründlich im Liebe "rumoren" und die Pariser Commune, welche auch in Deutschland ihre Schutzredner und Fürsprecher hat, war vielleicht nur eine leichtere Unverdaulichkeit gewesen an dem, was kommt.*

[1039] Obama 2014, not paginated.

[1040] Olmstead 1928, p. 277 US 479.

[1041] O'Reilly 1971, p. not paginated.

[1042] Orwell 2003, paraphrased from pp. 309-310.

[1043] Orwell 1968, p. 532.

[1044] Orwell 2003, p. 1.

[1045] Orwell 2003, p. 9.

[1046] Paine 1935, p. 167.

[1047] Paine 1935, p. 75.

[1048] Palmer 1918, p. 108.

[1049] Pankhurst 1913, p. 19.

[1050] Pankhurst 1914, p. 129.

[1051] Emergency Measure 1974, paraphrased from source text: *1974년 1월 8일에 제정된「대통령긴급조치」제1호는 대한민국 헌법을 부정·반대·왜곡 또는 비방하는 일체의 행위를 금하며, 대한민국 헌법의 개정 또는 폐지를 주장·발의·제안, 또는 청원하는 일체의 행위를 금하는 내용으로 구성되어*

있다. 또한 유언비어를 날조, 유포하는 일체의 행위를 금하고, 위에서 금한 행위를 권유·선동·선전하거나, 방송·보도·출판 기타 방법으로 이를 타인에게 알리는 일체의 언동을 금하였다. 이 조치에 위반한 자와 이 조치를 비방한 자는 법관의 영장없이 체포·구속·압수·수색하며 *15*년이하의 징역에 처하고, 이 경우에는 *15*년 이하의 자격정지를 병과할 수 있으며, 이 조치에 위반한 자와 이 조치를 비방한 자는 비상군법회의에서 심판, 처단한다. 이 조치는 *1974*년 *1*월 *8*일 *17*시부터 시행되었다.

1052 Freedom of Speech 2019, p. 31.

1053 Thucydides 2003, p. 396.

1054 Thucydides 2003, paraphrased from p. 397.

1055 Petronius 2005, paraphrased from p. 83. Source text: *Nunquam autem recte faciet, qui cito credit.*

1056 Phillips 1884, p. 235.

1057 Phillips 1852, p. 13.

1058 Piketty 2014, p. 2.

1059 Pinker 2009, not paginated.

1060 Pope Leo XIII 1901, article 9.

1061 Priestly 1949, p. 13.

1062 Prince 1944, p. 321 US 170.

1063 Ra 2012, timestamp 00:01:05.

1064 Rabelais 1913, p. 159. Source text: *Sapience n'entre point en âme malivole [sic], et science sans conscience n'est que ruine de l'âme.*

1065 Rand 1967, p. 138.

1066 Rao 1978, p. 4.

1067 Rao 1978, p. 48.

1068 Rao 1978, p. 32.

1069 Zauzmer 2016, not paginated. Note: In 2006 the Maryland legislature debated several bills to prohibit or endorse same-sex marriage: Senate Bills 262 and 690, House Bills 48 and 1637, and state law 2-201. The selected quote is taken from those debates.

1070 Rauf 2012, p. 49.

1071 Rawls 1971, p. 3.

1072 Rawls 1971, p. 210.

1073 Reagan 1987, not paginated.

1074 Robertson 2018, timestamp 00:31:30.

1075 Robertson 2018, timestamp 01:10:00.

1076 Roosevelt 1935, p. 8.

1077 Roosevelt 1949, not paginated.

[1078] Roosevelt 1938, p. 6.

[1079] Roosevelt 1908, not paginated.

[1080] Davis 1920, p. 646.

[1081] Rousseau 2003, p. 387.

[1082] Rousseau 2003, p. 390.

[1083] Rousseau 2003, p. 438.

[1084] Roy 1981, p. 73.

[1085] Rubin 2015, timestamp 00:00:45.

[1086] Vakil 2008, not paginated.

[1087] Russell 1950, p. 71.

[1088] Russell 1998, p. 28.

[1089] Ryan 1943, p. 141.

[1090] Sagan 1994b, not paginated.

[1091] Schurz 1913b, p. 258.

[1092] Schurz 1913a, p. 120.

[1093] Seneca the Younger 1953, epistle 104, pp. 210-211. Source text: *Non protest gratis constare libertas.*

[1094] Shelley 1915a, p. 17.

[1095] London 2016, not paginated.

[1096] Dawkins 2016, not paginated.

[1097] Smith 2003, paraphrased from p. 348.

[1098] Smith 1952, p. 300.

[1099] Smith 1972, p. 14.

[1100] Snyder 2017, not paginated.

[1101] Stobaeus 2016, p. 141. Source text: *συμβούλευε μή τά ήδιστα, άλλά τά βέλτιστα.*

[1102] Solzhenitsyn 1974, p. 38.

[1103] Sowell 1996, p. 118.

[1104] Spencer 1883, p. 94.

[1105] Spinelli 2017, p. 9. Source text: *La civiltà moderna ha posto come proprio fondamento il principio della libertà, secondo il quale l'uomo non deve essere un mero strumento altrui, ma un autonomo centro di vita.*

[1106] Stanislaw 1968, p. 4.

[1107] Stanley 1969, p. 394 US 565.

[1108] Stanton 1848, page 7. Hyphens added for emphasis.

[1109] Steinem 1986, p. 326.

[1110] Stephens 1866, p. 741.

[1111] Safire 1992, p. 72.

[1112] Stiglitz 2011, not paginated.

[1113] Stowe 1896, p. 523.

[1114] Strachey 1933, p. 138.

[1115] Tacitus 1956b, book 6, pp. 226-227. Source text: *Nam populi imperium iuxta libertatem, paucorum dominatio regiae libidini propior est.*

[1116] Tacitus 1956a, book 3, pp. 566-567. Source text: *Corruptissima re publica plurimae leges.*

[1117] Thatcher 1989, not paginated.

[1118] Thatcher 1991, not paginated.

[1119] Thoreau 1993, p. 18.

[1120] Thoreau 1993, p. 11.

[1121] Tocqueville 2003, p. 231.

[1122] Torcaso 1961, p. 367 US 495.

[1123] Treaty of Perpetual Peace 1797, article 11. Note: A revised treaty in 1805 omits the quoted text.

[1124] Trenchard 1720, p. 96.

[1125] Notes and Clippings 1889, p. 1.

[1126] Truman 1963, p. 238.

[1127] Bradford 1886, p. 29.

[1128] Tuchman 1980, p. 234.

[1129] Franklin's Motto 1811, p. 167. Source text: *Eripuit coelo fulmen sceptrumque tyrannis*. Also attributed to Marcus Manilius.

[1130] Freeman 2015, not paginated.

[1131] Universal Declaration of Human Rights 1948, article 21.

[1132] Virginia Assembly 1779, not paginated.

[1133] Voltaire 1922, p. 253. Source text: *Il vit combien il est dangereux d'avoir raison dans des choses où des hommes accrédités ont tort.*

[1134] Warburton 2009, p. 80.

[1135] Washington 1790, letter 289.

[1136] Motto 2017.

[1137] Weber 1946, paraphrased from p. 116.

[1138] Weber 1946, p. 115.

[1139] Weil 1978, p. 103.

[1140] Weiss 1890, p. 367.

[1141] Wells 1934, p. 618.

[1142] Wels 1933, not paginated. Source text: *Freiheit und Leben kann man uns nehmen, die Ehre nicht.*

[1143] White 1943, not paginated.

1144 Willis 2001, not paginated.

1145 Willson 1932, p. 22. First Amendment of the United States Constitution: *Congress shall make no law respecting an establishment of religion*.

1146 Wilson 1912, not paginated.

1147 Wilson 1919, p. 151.

1148 Wolfe 1980, p. 15.

1149 Wong 2019, not paginated.

1150 Wright 1972, p. 1210.

1151 Qin Yongmin 2018, not paginated.

1152 The World This Week 2019, p. 8.

1153 China's Newsrooms 2016, not paginated.

1154 Yousafzai 2013, not paginated.

1155 Zinn 2008, not paginated.

1156 Zola 1898, page 1. Source text: J'accuse.

1157 Zorach 1952, p. 343 US 325.

1158 Copeland 1999c, p. 882.

WORKS CITED

The bibliography is organized by author surname and publication date.

Abrams v United States. 1919. United States Supreme Court. Justice Oliver Wendell Holmes dissenting, citation 250 US 616, 10 November 1919.

Acton, John Dalberg. 1919a. *Historical Essays & Studies*. Edited by John Neville Figgis and Reginald Vere Laurence. London: MacMillan.

Acton, John Dalberg. 1919b. *The History of Freedom and Other Essays*. First published 1907. London: MacMillan.

Adams, John. 1780. *Constitution of the Commonwealth of Massachusetts*. Commonwealth of Massachusetts, ratified 25 October 1780.

Adams, John. 1788. *A Defence of the Constitutions of Government of the United States of America, Volume 3*. London: C. Dilly.

Adams, John. 1824. *The Trial of the British Soldiers of the 29th Regiment of Foot*. Trial held on 27 November 1770. Boston: William Emmons.

Adams, Douglas. 2002. *The Salmon of Doubt: Hitchhiking the Galaxy One Last Time*. New York: Ballantine.

Addison, Joseph. 1898. *The Spectator, Volume 3*. First published 1711. London: John C. Nimmo.

Adler, Alfred. 1979. *Religion and Individual Psychology*. First published 1933. Superiority and Social Interest: A Collection of Later Writings. Edited by Heinz L. Ansbacher and Rowena R. Ansbacher, third edition. New York: W. W. Norton.

Adorno, Theodor. 1970. *Minima Moralia: Reflexionen aus dem beschädigten Leben*. First published 1951. Frankfurt: Suhrkamp.

Aitkenhead, Decca. 2016. Prof Brian Cox: Being Anti-Expert - That's the Way Back to the Cave. *The Guardian*. 2 July 2016.

Al-Khalili, Jim. 2011. *The House of Wisdom: How Arabic Science Saved*

Ancient Knowledge and Gave Us the Renaissance. New York: Penguin.

Alexander, David. 1991. The Humanist Interview with Gene Roddenberry. *The Humanist*. March/April 1991.

Allen, Ethan. 1836. *Reason, the Only Oracle of Man: Or a Compendious System of Natural Religion*. New York: G. W. & A. J. Matsell.

Allen, Woody. 1969. My Philosophy. *The New Yorker*. 19 December 1969.

Allen, Woody. 1980. *Stardust Memories*. Voiced as character Sandy Bates. Metro Goldwyn Mayer. [video]

Allen, Steve. 1990. *On the Bible, Religion and Morality*. New York: Prometheus.

Allen, Steve. 1994. *Reflections*. New York: Prometheus.

Amiel, Henri-Frédéric. 1893. *Amiel's Journal: The Journal Intime of Henri-Frédéric Amiel*. Translated by Mrs. Humphry Ward. London: MacMillan.

Amorth, Gabriele. 2002. *An Exorcist: More Stories*. First published 1992. San Francisco: Ignatius Press.

Andrews, Robert. 1989. *The Concise Columbia Dictionary of Quotations*. New York: Columbia University Press.

Angelou, Maya. 1980. *I Know Why the Caged Bird Sings*. First published 1970. New York: Bantam.

Arabi, Soheil. 2017. The Life of Soheil Arabi Is in Danger. Let's Be His Voice. *Facebook*. 30 September 2017.

Arago, François. 1857. *The Biographies of Distinguished Scientific Men*. Translated by W. H. Smith, Baden Powell, and Robert Grant. London: Longman, Brown, Green, Longmans & Roberts.

Archer, William. 1917. *God and Mr. Wells*. New York: Alfred A. Knopf.

Arendt, Hannah. 1962. *The Origins of Totalitarianism*. Seventh printing. New York: Meridian.

Aristotle. 2003a. *Nicomachean Ethics*. First published circa 350 BCE. Great Books of the Western World, Volume 8. Edited by Mortimer J. Adler, translated by W. D. Ross, seventh printing. Chicago: Encyclopædia Britannica.

Aristotle. 2003b. *Politics*. First published circa 350 BCE. Great Books of the Western World, Volume 8. Edited by Mortimer J. Adler, translated by Benjamin Jowett, seventh printing. Chicago: Encyclopædia Britannica.

Aristotle. 2003c. *Metaphysics*. First published circa 350 BCE. Great Books of the Western World, Volume 7. Edited by Mortimer J. Adler, translated by G. R. G. Mure, seventh printing. Chicago: Encyclopædia

Britannica.

Arnold, Mathew. 1888. *Essays in Criticism: Second Series*. London: MacMillan.

Arnold, Mathew. 1903a. *God and the Bible: A Sequel to Literature and Dogma*. First published 1875. London: Smith, Elder.

Arnold, Mathew. 1903b. *Literature and Dogma: An Essay Towards a Better Apprehension of the Bible*. First published 1873. London: Watts.

Ashton, John, editor. 2009. *In Six Days: Why Fifty Scientists Choose to Believe in Creation*. Seventh printing. Green Forest: Master.

Asimov, Isaac. 1980. A Cult of Ignorance. *Newsweek*. 21 January 1980.

Asimov, Isaac. 1981. The Threat of Creationism. *The New York Times*. 14 June 1981.

Asimov, Isaac. 1983. *The Roving Mind*. Oxford: Oxford University Press.

Asimov, Isaac. 1988. *Isaac Asimov - Part II*. Bill Moyers' World of Ideas. Show #130, 21 October 1988.

Asimov, Isaac. 2004. *Foundation*. First published 1951. New York: Bantam.

Atkins, Peter. 1996. Professor Says Science Rules Out Belief in God. *The Daily Telegraph*. Reported by Roger Highfield, 11 September 1996.

Atkins, Peter. 2000. Science and Religion: Rack or Featherbed - The Uncomfortable Supremacy of Science. *Science Progress*. Volume 83, number 1, Millennium Issue 2000.

Attenborough, David. 1979. *Life of Earth: A Natural History*. Boston: Little, Brown.

Atwood, Margaret. 1990. Margaret Atwood Reflects on a Hit. *The New York Times*. The Arts, 14 April 1990.

Augustine of Hippo. 1874. *Lectures or Tractates of the Gospel According to St. John, Volume 2*. First published circa 415. The Works of Aurelius Augustine, Bishop of Hippo, Volume 11. Edited by Marcus Dods, translated by James Innes. Edinburgh: T. & T. Clark.

Augustine of Hippo. 1892. *On Christine Doctrine*. First published 397-426. The Works of Aurelius Augustine, Bishop of Hippo, Volulme 9. Edited by Marcus Dods. Edinburgh: T. & T. Clark.

Aurelius, Marcus. 2003. *The Meditations of Marcus Aurelius*. First published circa 200-400 CE. Great Books of the Western World, Volume 11. Edited by Mortimer J. Adler, translated by George Long, seventh printing. Chicago: Encyclopædia Britannica.

Austin, David. 1983. The Trouble with Thinking Backwards. *New Scientist*. Cartoon accompanying article by Ralph Estling, volume 80, number 1360, 2 June 1983

Avni, Benny. 1989. An Interview with Abbie Hoffman. *Tikkun*. Volume 4, number 4, July/August 1989, pp. 15-19.

Ayala, Francisco J. 1974. Biological Evolution: Natural Selection or Random Walk: Electrophoretic Analysis of the Genetic Constitutions of Four Drosophila Species Tends to Refute the Neutrality Theory of Protein Evolution. *American Scientist*. Volume 62, number 6, November-December 1974, pp. 692-701.

Ayer, A. J. 1952. *Language, Truth and Logic*. New York: Dover.

Ayer, A. J., editor. 1968. *The Humanist Outlook*. London: Pemberton.

Bacon, Francis. 1657. *A Proposition to His Majesty, by Sir Francis Bacon, Knight, His Majesties Attorney Generall; and One of His Privy Counsell; Touching the Compiling, and Amendment, of the Lawes of England*. First published circa 1620. Resuscitatio, Or, Bringing Into Publick Light Severall Pieces, of the Works, Civil, Historical, Philosophical, & Theological, Hitherto Sleeping, of the Right Honourable Francis Bacon. London: Sarah Griffin.

Bacon, Francis. 1902. *Novum Organum*. First published 1620. Edited by Joseph Devey. New York: P. F. Collier.

Bacon, Francis. 1908. *Of Vicissitudes of Things*. First published 1625. The Essays Of Francis Bacon. Edited by Mary Augusta Scott. New York: Charles Scribner's Sons.

Bacon, Francis. 2003. *Novum Organum*. First published 1620. Great Books of the Western World, Volume 28. Edited by Mortimer J. Adler, seventh printing. Chicago: Encyclopædia Britannica.

Badawi, Raif. 2015. *Twitter (raifbadawiquote)*. 15 January 2015, 09:16 a.m. USA CST.

Baker, Russell. 1988. The Real Thing. *The New York Times*. 21 December 1988.

Bakunin, Mikhail. 1895. *God and the State*. Translated by Benjamin R. Tucker. New York: Benjamin R. Tucker.

Baldwin, James. 1962. Letter from a Region in my Mind. *The New Yorker*. 17 November 1962.

Baldwin, James. 2007. *No Name in the Street*. First published 1972. New York: Vintage.

Baldwin, James. 2011. The Crusade of Indignation. First published 1956. *The Nation*. 7 July 1956.

Ball, Philip. 2018. *Beyond Weird: Why Everything You Thought You Knew About Quantum Physics Is Different*. Chicago: University of Chicago Press.

Ballou, Martin. 1881. *Treasury Of Thought*. Boston: Houghton Mifflin.

Ballou, Martin. 1886. *Edge-Tools Of Speech*. Cambridge: Riverside Press.

Balzac, Honoré de. 1892. *Sur Catherine de Médicis*. Paris: Calmann Lévy.

Barker, Dan. 2006. *Losing Faith in Faith: From Preacher to Atheist*. First published 1992. Madison: Freedom From Religion Foundation.

Barker, Dan. 2008. *Godless: How an Evangelical Preacher Became One of America's Leading Atheists*. Berkeley: Ulysses Press.

Barnes, Ernest William. 1933. *Scientific Theory and Religion: The World Described by Science and Its Spiritual Interpretation*. From the Gifford lectures at Aberdeen, 1927-1929. New York: MacMillan.

Barnhill, John Basil. 1914. *Indictment of Socialism No. 3*. Barnhill-Tichenor Debate on Socialism. Rip-Saw Library, Number 101. Saint Louis: National Rip-Saw.

Barry, Dave. 1998. *Dave Barry Turns 50*. New York: Ballantine.

Barsky, Robert F. 1997. *Chomsky: A Life of Dissent*. Toronto: ECW Press.

Baruch, Bernard. 1957. *Baruch: My Own Story*. Cutchogue: Buccaneer.

Baskerville, John. 1907. *John Baskerville: A Memoir*. Cambridge: University Press.

Bastiat, Frédéric. 1854. *Sophismes Économiques*. Oeuvres Complètes de Frédéric Bastiat, Volume 4. Paris: Guillaumin.

Bates, M. Searle. 1945. *Religious Liberty: An Inquiry*. New York: International Missionary Council.

Batson, C. Daniel, Patricia Schoenrade, and W. Larry Ventis. 1993. *Religion and the Individual: A Social-Psychological Perspective*. New York: Oxford University Press.

Bayle, Pierre. 1820. *Dictionnaire Historique et Critique de Pierre Bayle, Volume 1*. First published 1679. Paris: Desoer.

Bayle, Pierre. 1826. *An Historical and Critical Dictionary, Volume 3*. First published 1679. London: Hunt & Clarke.

Bayle, Pierre. 1911. *Pensées Diverses sur la Comète, Volume 1*. First published 1682. Paris: Édouard Cornély.

Beauvoir, Simone de. 1976. *Le Deuxième Sexe*. First published 1949. Paris: Éditions Gallimard.

Beccaria, Cesare. 1775. *An Essay on Crimes and Punishments*. First published 1764. Fourth edition. London: F. Newberry.

Beit-Hallahami, Benjamin, and Michael Argyle. 1997. *The Psychology of Religious Behaviour, Belief and Experience*. London: Routledge.

Bell Burnell, Jocelyn. 2010. *Beautiful Minds, Series 1*. 7 April 2010. Aired by British Broadcasting Corporation (BBC). [video]

Bellamy, Edward. 1913. *Equality*. New York: D. Appleton.

Benson, Herbert. 2005. Study of the Therapeutic Effects of Intercessory Prayer (STEP) in Cardiac Bypass Patients: A Multicenter Randomized Trial of Uncertainty and Certainty of Receiving Intercessory Prayer. *American Heart Journal*. Volume 151, number 4, May 2005, pp. 934-942.

Bentham, Jeremy. 1834. *Deontology: Or The Science of Morality, Volume 1*. Edited by John Bowring. London: Longman, Rees, Orme, Brown, Green & Longman.

Bentham, Jeremy. 1969. *A Fragment on Government*. A Bentham Reader. Edited by Mary Peter Mack. New York: Pegasus.

Bentley, Eric, editor. 1966. *Galileo*. First published 1948. New York: Glove Press.

Berkeley, George. 1747. *Siris: A Chain of Philosophical Reflexions and Inquiries Concerning theVirtues of Tarwater, and Divers Other Subjects Connected Together and Arising One From Another*. First published 1744. London: C. Davies.

Berkeley, George. 1906. *Three Dialogues Between Hylas and Philonous*. First published 1713. Chicago: Open Court.

Berlin, Isaiah. 2002. *Notes on Prejudice*. First published 2001. Liberty: Incorporating Four Essays on Liberty. Edited by Henry Hardy. Oxford: Oxford University Press.

Bernal, J.D. 1939. *The Social Function Of Science*. London: Routledge.

Bernanos, Georges. 1968. *Why Freedom?* First published 1953. The Last Essays of Georges Bernanos. Translated by Joan and Barry Ulanov. New York: Greenwood Press.

Bernanos, Georges. 1971. *Nous Autres Français*. First published 1939. Essais et Écrits de Combat, Volume 1. Edited by Michel Estève. Paris: Gallimard.

Bernard, Abbot of Arbroath. 1320. *The Declaration of Arbroath*. National Records of Scotland.

Bharara, Preet. 2019. *Doing Justice: A Prosecutor's Thoughts on Crime, Punishment and the Rule of Law*. New York: Alfred A. Knopf.

Bhutto, Benazir. 2007. This Is My Oath: I Will Never Give in to Tyranny. *The Times (London)*. 18 October 2017.

Bierce, Ambrose. 1911. *Epigrams*. The Collected Works of Ambrose Bierce, Volume 3. New York: Neale.

Bierce, Ambrose. 1993. *The Devil's Dictionary*. First published 1911. New York: Dover Publications.

Biskupski, M. B., and James S. Paula, editors. 1990. *The General Confederation of Warsaw 1573*. Signed 28 January 1573. Polish Democratic Thought From the Renaissance to the Great Emigration: Essays and Documents. New York: Columbia University Press (East European Monographs).

Blackstone, William. 1768. *Commentaries on the Laws of England, Book 1*. Third edition. Oxford: Clarendon Press.

Boileau, Nicolas. 1872. *L'Art Poétique*. First published 1674. Oeuvres Complètes de Boileau, Volume 2. Paris: Garnier Frères.

Boltzmann, Ludwig. 1974. *On a Thesis of Schopenhauer's*. First published 1905. Theoretical Physics and Philosophical Problems. Translated by Paul Foulkes. Dordrecht: D. Reidel.

Bonner, Hypatia Bradlaugh. 1919. *Christianity and Conduct, Or, The Influence of Religious Beliefs on Morals*. London: Watts.

Boorstin, Daniel J. 1962. *The Image*. New York: Harper Colophon.

Boosler, Elayne. 1987. *Broadway Baby*. Aired by Showtime. [video]

Bossuet, Jacques-Bénigne. 1821. *Politique Tirée des Propres Paroles de l'Écriture Sainte*. First published 1709. Paris: Monarchique.

Bowen, E. J. 1967. Sir Cyril Hinshelwood. *Chemistry in Britain*. Volume 3, pp. 534-536.

Boyer, Pascal. 2001. *Religion Explained: The Evolutionary Origins of Religious Thought*. New York: Basic.

Bradbury, Ray. 1979. *The Martian Chronicles*. First published 1950. New York: Bantam.

Bradford, Sarah. 1886. *Harriet: The Moses of Her People*. New York: G. R. Lockwood.

Bradlaugh, Charles. 1870. *A Plea for Atheism*. London: Austin.

Bradlaugh, Charles. 1889. *Humanity's Gain from Unbelief*. London: Freethought.

Bradley, Ian, editor. 1996. *Princes Ida*. First published 1884. The Complete Annotated Gilbert and Sullivan. Oxford: Oxford University Press.

Bragg, Melvyn. *In Our Time*. British Broadcasting Corporation (BBC). [audio podcast]

Brandon, Barbara. 1986. *The Passion of Ayn Rand*. New York: Doubleday.

Bronowski, Jacob. 1956. Science and Human Values. New York: Harper & Brothers.

Brooks, David Marshall. 1933. *The Necessity of Atheism*. New York: Freethought Press.

Brown, J. A. C. 1963. *Techniques of Persuasion: From Propaganda to*

Brainwashing. Baltimore: Penguin.

Browne, Thomas. 1907. *Hydriotaphia: Urne-Buriall, or a Discourse of the Sepulchrall Urnes Lately Found in Norfolk*. First published 1658. Cambridge: Riverside Press.

Bruñel, Luis. 1983. *Mi Último Suspiro*. First published 1982. Translated by Ana Maria de la Fuente. Barcelona: Plaza & Janes.

Bryan, William Jennings. 1909. *Speech on the Jury System*. First published 1890. Speeches Of William Jennings Bryan, Volume 1. New York: Funk & Wagnalls.

Buchan, John. 1941. *The Law*. Canadian Occasions. Address to Law Society of Upper Canada, Toronto, 21 February 1936. Toronto: Musson.

Buck, Pearl S. 1938. America's Medieval Women. *Harper's Magazine*. Volume 177, August 1938, pp. 225-232.

Buckle, Henry Thomas. 1929. *History of Civilization in England, Volume 1*. First published 1857. New York: Appleton.

Buckley, William F., Jr. 2004. Free Weeds. *National Review*. 29 June 2004.

Bulmer-Thomas, Ivor. 1949. *The Socialist Tragedy*. Second edition. London: Latimer House.

Burbank, Luther. 1926. *Why I Am an Infidel*. Little Blue Book 1020. Edited by E. Haldeman-Julius. Girard: Haldeman-Julius.

Burke, Edmund. 1782. *Observations on a Late State of the Nation*. Fifth edition. London: J. Dodsley.

Burnet, John. 1908. *Early Greek Philosophy*. First published circa 1892. Second edition. London: Adam and Charles Black.

Burns, Robert. 1896. *The Rights of Woman*. First published 1792. The Poetry of Robert Burns, Volume 2. Edited by William Ernest Henley and Thomas F. Henderson. Edinburgh: T. C. and E. C. Jack.

Burroughs, John. 1900. *The Light of Day: Religious Discussions and Criticisms from the Naturalist's Point of View*. Boston: Houghton Mifflin.

Burtt, Edwin Arthur. 1925. *The Metaphysical Foundations of Modern Physical Science: A Historial and Critical Essay*. London: Kegan Paul, Trench, Trubner.

Bury, J. B. 1913. *A History of Freedom of Thought*. New York: Henry Holt.

Butler, Samuel. 1921. *The Note-Books of Samuel Butler*. First published 1917. New York: E. P. Dutton.

Butler, Samuel. 1986. *The Way of All Flesh*. First published 1903. Edited by James Cochrane. London: Penguin.

Butt, Riazat. 2009. US Teacher Broke Law by Describing Creationism as Superstitious Nonsense. *The Guardian*. 6 May 2009.

Cabell, James Branch. 1917. *The Cream of the Jest*. New York: Modern Library.

Caesarius of Heisterbach. 1591. *De Daemonibus*. First published circa 1220. Illustrium Miraculorum et Mistoriarum Memorabilium, Volume 12. Cologne: Birckmannica.

Calder, Nigel. 1970. *Violent Universe: An Eyewitness Account of the New Astronomy*. New York: Viking Press.

Calhoun, John C. 1848. *Conquest of Mexico: Speech of Mr. Calhoun of South Carolina in the Senate of the United States, 4 January 1848*. Appendix to the Congressional Globe. 30th Congress, 1st Session, 6 December 1847 to 14 August 1848, series 4, pp. 49-53.

Calvin, John. 1845a. *The Institutes of the Christian Religion, Volume 1*. First published 1536. Translated by Henry Beveridge. Edinburgh: Calvin Translation Society.

Calvin, John. 1845b. *The Institutes of the Christian Religion, Volume 2*. First published 1536. Translated by Henry Beveridge. Edinburgh: Calvin Translation Society.

Campbell, Lewis, and William Garnett. 1882. *The Life of James Clerk Maxwell: With a Selection From His Correspondence and Occasional Writings and a Sketch of His Contributions to Science*. London: MacMillan.

Campbell, Joseph. 1972. *Myths To Live By*. New York: Viking Press.

Campbell, Neil A., Jane B. Reece, and Lawrence G. Mitchell. 1999. *Biology*. First published 1987. Fifth edition. Menlo Park: Benjamin Cummings.

Camus, Albert. 1947. *La Peste*. Paris: Éditions Gallimard.

Carlin, George. 1997. *Brain Droppings*. New York: Hyperion.

Carlin, George. 2004. *When Will Jesus Bring the Pork Chops?* New York: Hyperion.

Carter, Angela. 1985. The Company of Angela Carter: An Interview. *Marxism Today*. January 1985, pp. 20-22.

Cathcart, Thomas, and Daniel Klein. 2008. *Plato and a Platypus Walk Into a Bar: Understanding Philosophy Through Jokes*. First published 2007. New York: Penguin.

Cerf, Bennett. 1945. *Try and Stop Me: A Collection of Anecdotes and Stories, Mostly Humorous*. New York: Simon & Schuster.

Chandler, Raymond. 1976. *The Notebooks of Raymond Chandler, and English Summer: A Gothic Romance*. Edited by Frank MacShane.

New York: Ecco Press.

Chartier, Émile. 1920. *Les Propos d'Alain, Volume 1*. Paris: Nouvelle Revue Française.

Châtelet, Émilie du. 1740. *Institutions de Physique*. Paris: Prault.

Chesnut, Mary Boykin. 2006. *A Diary From Dixie*. First published 1905. Diary entry from 2 September 1864. New York: Barnes & Noble.

Chesterton, G. K. 1908. *Heretics*. London: John Lane.

China's Newsrooms Serve as President Xi's Loyal Mouthpieces. 2016. The Post's View: Opinions of the Editorial Board. *Washington Post*. 23 February 2016.

Chisholm, Shirley. 1971. *Unbought and Unbossed*. New York: Avon.

Chomsky, Noam. 2003. *The Common Good*. Interview by David Barsamian, fifth printing. Tucson: Odonian Press.

Churchill, Winston. 1947. House of Commons: Parliament Bill. *Hansard*. Volume 444, 11 November 1947.

Cicero. 1960. *On Invention (De Inventione)*. First published 85 BCE. Loeb Classical Library, LCL 386. Translated by H. M. Hubbell. Cambridge: Harvard University Press.

Cicero. 2006. *On Duties (De Officils)*. First published 44 BCE. A Loeb Classical Library Reader. Translated by Walter Miller. Cambridge: Harvard University Press.

Clark, Arthur C. 1980. *Science and Spirituality*. Voices From The Sky. New York: Pocket.

Cleyre, Voltairine De. 2004. *Sex Slavery.* First published 1890. The Voltairine De Cleyre Reader. Oakland: AK Press.

Clifford, William Kingdon. 1879a. *On Some of the Conditions of Mental Development*. Lectures and Essays, Volume 1. Edited by Leslie Stephen and Frederick Pollock. London: MacMillan.

Clifford, William Kingdon. 1879b. *On the Aims and Instruments of Scientific Thought*. Lectures and Essays, Volume 1. Edited by Leslie Stephen and Frederick Pollock. London: MacMillan.

Coffin, Henry Sloan. 1935. *Address by Henry Sloane Coffin at the Memorial Service for Henry Fairfield Osborn, at St. Bartholmew's Church, New York, December 18, 1935*. Pamphlets on Biography, Kofoid Collection, Volume 33, University of California. [Also cited in *Supplement to Natural History*, Volume 37, number 2, February 1936]

Cohen, Morris R. 1933. *The Dark Side of Religion*. Religion Today: A Challenging Enigma. Edited by Arthur L. Swift. New York: Whittlesey House.

Cohen, Morris R., and Ernest Nagel. 1955. *An Introduction to Logic and Scientific Method*. First published 1934. London: Routledge & Kegan Paul.

Cohen, Chapman. 1980. *The Devil*. An Anthology of Atheism and Rationalism. Edited by Gordon Stein. Amherst: Prometheus.

Cohen, I. Bernard. 1985. *The Birth of a New Physics: Revised and Updated*. First published 1960. New York: W. W. Norton.

Coleridge, Taylor Coleridge. 1836. *Letters, Conversations and Recollections of S. T. Coleridge, Volume 1*. London: Edward Moxon.

Coleridge, Ernest Hartely, editor. 1895. *Anima Poetæ: From the Unpublished Books of Samuel Taylor Coleridge*. London: William Heinemann.

Collingwood, Robin. 1947. *The New Leviathan or Man, Society, Civilization and Barbarism*. First published 1840. Oxford: Clarendon Press.

Collins, Anthony. 1713. *A Discourse of Free-Thinking*. London: Publisher unknown.

Colman, Lucy. 1891. *Reminiscences*. Buffalo: H. L. Green.

Colton, Charles. 1825. *Lacon; Or Many Things in Few Words; Addressed to Those Who Think, Volume 1*. First published 1820. London: Longman, Hurst, Rees, Orme, Brown & Green.

Colton, Charles. 1829. *Lacon; Or Many Things in Few Words; Addressed to Those Who Think, Volume 2*. First published 1822. London: Longman, Hurst, Rees, Orme, Brown & Green.

The Commonwealth of Massachusetts. 2019. *General Laws, Part IV, Title 1, Chapter 272, Section 36*. The 191st General Court of the Commonwealth of Massachusetts.

Condorcet, Marquis de. 1847. *Sur l'Admission des Femmes au Droite de Cité*. First published 1790. Œuvres de Condorcet, Volume 10. Paris: Firmin Didot Frères.

Conklin, Edwin Grant. 1937. Science and Ethics. *Science*. Volume 86, number, 2244, 31 December 1937, pp. 595-603.

Connor, Steve. 2011. For the Love of God: Scientists in Uproar at £1m Religion Prize. *The Independent*. 7 April 2011.

Conway, Moncure. 1884. *Apologia*. Farewell Discourses Delivered at South Place Chapel, Finsbury, London. London: E. W. Allen.

Copeland, Lewis, et al., editors. 1999a. *Address at Angostura, 15 February 1819*. The World's Great Speeches: 292 Speeches from Pericles to Mandela. First published 1942. Fourth edition. Mineola: Dover.

Copeland, Lewis, et al., editors. 1999b. *To Workingmen and Soldiers,*

March 1917. The World's Great Speeches: 292 Speeches from Pericles to Mandela. First published 1942. Fourth edition. Mineola: Dover.

Copeland, Lewis, et al., editors. 1999c. *Statement of Support for Student Pro-Democracy Movement, 16 May 1989*. The World's Great Speeches: 292 Speeches from Pericles to Mandela. First published 1942. Fourth edition. Mineola: Dover.

Copernicus, Nicolaus. 1566. *De Revolutionibus Orbium Coelestium*. First published 1543. Basel: Henricus Petrus.

Cousins, Norman. 1953. *Who Speaks for Man?* New York: MacMillan.

Coward, Noël. 1979. *Blithe Spirit*. First published 1941. Private Lives, Blithe Spirit, Hay Fever. New York: Grove Weidenfeld.

Crisp, Quentin. 2000. *How to Become a Virgin*. First published 1981. The Naked Civil Servant; How to Become a Virgin; Resident Alien. New York: Quality Paperback Book Club.

Cromwell, Oliver. 1904. *The Letters and Speeches of Oliver Cromwell, Volume 2*. Speech delivered 20 April 1653, edited by S. C. Lomas. London: Methuen.

Curie, Eve. 1947. *Madame Curie*. First published 1938. Translated by Vincent Sheean. London: William Heinemann.

Cuvier, Georges. 1810. *Rapport Historique sur les Progrés des Sciences Naturelles Depuis 1789, et sur Leur État Actuel*. Paris: Imprimerie Impériale.

Darrow, Clarence. 1920. *Argument of Clarence Darrow in the Case of the Communist Labor Party in the Criminal Court, Chicago*. Chicago: Charles Keer.

Darrow, Clarence. 1929. *Why I Am an Agnostic: Including Expressions of Faith From a Protestant, a Catholic and a Jew*. Little Blue Book 1500. Edited by E. Haldeman-Julius. Girard: Haldeman-Julius.

Darwin, Charles. 2003. *The Descent of Man, and Selection in Relation to Sex*. First published 1871. Great Books of the Western World, Volume 49. Edited by Mortimer J. Adler, seventh printing. Chicago: Encyclopædia Britannica.

Davis, Jefferson. 1881. *The Rise and Fall of the Confederate Government, Volume 2*. New York: D. Appleton.

Davis, Philip, and Bertha Schwartz, editors. 1920. *Address to the Knights of Columbus. Immigration and Americanization: Selected Readings*. Delivered 12 October 1915. Boston: Ginn.

Dawkins, Richard. 1996. *Science, Delusion and the Appetite for Wonder*.

The Richard Dimbleby Lecture. 12 November 1996. Aired by British Broadcasting Corporation (BBC). [video]

Dawkins, Richard. 2009. *The Greatest Show on Earth: The Evidence for Evolution*. New York: Free Press.

Dawkins, Richard. 2016. *The Selfish Gene: Fortieth Anniversary Edition*. First published 1976. Fortieth Anniversary Edition. Oxford: Oxford University Press.

Debate on Modern Abolitionism. 1836. *General Conference of the Methodist Episcopal Church*. May 1836. Cincinnati: Ohio Anti-Slavery Society.

Debs, Eugene. 1994. *The Canton, Ohio, Speech, 16 June 1918*. Eugene V. Debs Speaks. First published 1970. New York: Pathfinder.

Delay, Tom. 1999. Consequences for Juvenile Offenders Act of 1999. *Congressional Record*. 106th Congress, first session, volume 145, number 85, 16 June 1999.

Dell, Diana, editor. 2000. *Memorable Quotations: Humorists, Wits and Satirists of the Past*. San Jose: Writers Club Press.

Deloria, Vine, Jr. 1969. *Custer Died for Your Sins; An Indian Manifesto*. New York: Avon.

Demosthenes. 1930. *Olynthiacs*. First published 349 BCE. Loeb Classical Library, LCL 238. Translated by J. H. Vince. Cambridge: Harvard University Press.

Dennert, Eberhard. 1903. *Vom Sterbelager des Darwinismus: Ein Bericht*. Stuttgart: Max Kielmann.

Dennett, Daniel. 1995. *Darwin's Dangerous Idea: Evolution and the Meanings of Life*. New York: Simon & Schuster.

Dennett, Daniel. 2017. *From Bacteria to Bach and Back: The Evolution of Minds*. New York: W. W. Norton.

Denny, George V., Jr (host). 1938. *Bulletin of America's Town Hall on the Air*. Quoted by Wendell L. Willkie, volume 3, number 10, 6 January 1938. New York: NBC Blue Network. [audio transcript]

Descartes, René. 1910. *Discours de la Méthode*. First published 1637. Paris: Hachette.

Descartes, René. 1911. *The Principles of Philosophy*. The Philosophical Works of Descartes, Volume 1. Translated by Elizabeth S. Haldane and G. R. T. Ross. Cambridge: Cambridge University Press.

Deschner, Karlheinz. 1996. *Bissige Aphorismen*. Reinbek bei Hamburg: Rowohlt.

Deschner, Karlheinz. 2013. *Kriminalgeschichte des Christentums, Volume 3*.

First published 1990. Fourth edition. Reinbek bei Hamburg: Rowohlt.

Dewey, John. 1910a. *How We Think*. Boston: D. C. Heath.

Dewey, John. 1910b. *The Influence of Darwin on Philosophy, and Other Essays in Contemporary Thought*. New York: Henry Holt.

Dewey, John. 1929. *The Quest for Certainty: A Study of the Relation of Knowledge and Action*. London: George Allen & Unwin Ltd.

Dick, Philip K. 1996. *How To Build A Universe That Doesn't Fall Apart Two Days Later*. The Shifting Realities of Philip K. Dick: Selected Literary and Philosophical Writings. Edited by Lawrence Sutin. New York: Vintage.

Dickens, Charles. 1998. *Great Expectations*. First published 1861. Oxford: Oxford University Press.

Diderot, Denis, and Jean le Rond d'Alembert. 1779. *Encyclopédie, ou Dictionnaire Raisonné des Sciences, des Arts et des Métiers, par une Société de Gens de Lettres, Volume 25*. First published 1755. Genève: Jean- Léonard Pellet.

Diderot, Denis. 1819. *Thoughts on Religion*. London: Carlile.

Diderot, Denis. 1821a. *Essai sur le Mérite de la Vertu*. Œuvres de Denis Diderot, Volume 1. First published 1745. Paris: J. L. J. Brière.

Diderot, Denis. 1821b. *Pensées Philosophiques*. Œuvres de Denis Diderot, Volume 1. First published 1746. Paris: J. L. J. Brière.

Dillahunty, Matt. 2008. *The Atheist Experience*. Episode 578, 9 November 2008. [video]

Dillard, Annie. 1998. *Pilgrim At Tinker Creek*. First published 1985. New York: Perennial Classics.

Dilliard, Irving. 1952. *The Spirit of Liberty*. New York: Alfred A. Knopf.

Diodorus of Sicily. 1933. *The Library of History, Volume 1*. First published circa 30 BCE. Loeb Classical Library, LCL 279. Translated by C. H. Oldfather. Cambridge: Harvard University Press.

Disraeli, Benjamin. 1879. *Lothair*. London: Longmans, Green.

Dobzhansky, Theodosius. 1945. Evolution, Creation and Science. *The American Naturalist*. Volume 79, number 780, January-February 1945.

Doren, Carl Van. 1926. *Why I Am an Unbeliever. Twelve Modern Apostles and Their Creeds*. New York: Duffield.

Douglass, Frederick. 1847. *Narrative of the Life of Frederick Douglass, an American Slave*. Boston: Anti-Slavery Office.

Douglass, Frederick. 1857. *Two Speeches by Frederick Douglass; One on West India Emancipation, Delivered at Canandaigua, Aug. 4th, and the Other on the Dred Scott Decision, Delivered in New York, on the*

Occasion of the Anniversary of the American Abolition Society, May, 1857. Rochester: C. P. Dewey.

Douglass, Frederick. 1893. *Haiti and the Haitian People: An Address Delivered in Chicago, Illinois on 2 January 1893*. The Frederick Douglass Papers, Series One, Speeches, Debates and Interviews, Volume 5: 1881-1895. Edited by John Blassingame and John McKivigan. New Haven: Yale University Press.

Doyle, Arthur Conan. 1891. The Adventures of Sherlock Holmes, Adventure IV, The Boscombe Valley Mystery. *The Strand Magazine*. Volume 2, number 10, July to December 1891.

Draper, John William. 1880. *History of the Conflict Between Religion and Science*. First published 1874. Fourteenth edition. London: C. Kegan Paul.

Drennan, Robert E., editor. 1995. *The Algonquin Wits*. First published 1968. New York: Carol.

Duclaux, Émile. 1896. *Pasteur: Histoire d'un Esprit*. Sceau: Imprimerie Charaire.

Duhem, Pierre. 1906. *La Théorie Physique: Son Objet et Sa Structure. Bibliothèque de Philosophie Expérimentale, Volume 11*. Edited by E. Peillaube. Paris: Chevalier & Rivière.

Dumas, Alexandre. 1880. *Le Vicomte de Bragelonne: Ou, Dix Ans Plus Tard; Complément des Trois Mousquetaires et de Vingt Ans Après, Volume 1*. First published 1847. Paris: Calmann Lévy.

Durant, Will, and Ariel Durant. 2012. *The Lessons of History*. New York: Simon & Schuster.

Dylan, Bob, and Sam Shepard. 1986. *Brownsville Girl*. Knocked Out Girl. Columbia Records. [song lyric]

Eade, Charles. 1945. *The Crisis in Greece, 8 December 1944*. The Dawn of Liberation: War Speeches by the Right Hon. Winston S. Churchill, C. H., M.P. London: Cassell.

Eddington, Arthur. 1920. *Space Time and Gravitation: An Outline of the General Relativity Theory*. Cambridge: Cambridge University Press.

Eddington, Arthur. 1929. *Science and the Unseen World*. New York: MacMillan.

Einstein, Albert. 1919. Induktion und Deduktion in der Physik. *Berliner Tageblatt*. Number 617, 4th supplement, 25 December 1919.

Einstein, Albert. 1922. *Letter to his Sister, 12 August 1922*. Kedem Auction House: Rare and Important Items. Auction 63, lot 176, auction held 13 November 2018.

Einstein, Albert. 1950. *The Laws of Science and the Laws of Ethics*. Out of My Later Years. New York: Philosophical Library.

Einstein, Albert. 1954. *Letter to Erik Gutkind, 3 January 1954*. Christie's: Albert Einstein - The God Letter. Auction 16393, lot 1, auction held 4 December 2018.

Einstein, Albert. 1982. *Ideas and Opinions*. First published 1954. Translated by Sonja Bargman, edited by Carl Selig. New York: Crown.

Einstein, Albert. 1987. *Letter to Jost Winteler, 8 July 1901*. The Collected Papers of Albert Einstein, Volume 1, The Early Years, 1879-1902. Edited by John Stachel, David C. Cassidy, and Robert Schulmann. Princeton: Princeton University Press.

Eiseley, Loren C. 1975. *The Firmament of Time*. First published 1960. New York: Antheneum.

Eldridge, Niles. 1985. *Time Frames: The Re-Thinking of Darwinian Evolution and the Theory of Punctuated Equilibria*. Princeton: Princeton University Press.

Eliot, George. 1883. *Evangelical Teaching: Dr. Cumming*. First published 1855. Essays and Leaves From a Note-Book. New York: Harper & Brothers.

Elliot, Jonathan, editor. 1861. *The Debates of the Adoption of the Federal Constitution in the Convention Held at Philadelphia in 1787*. The Debates in the Several State Conventions on the Adoption of the Federal Constitution as Recommended by the General Convention at Philadelphia in 1787, Volume 5. Philadelphia: J. B. Lippincott.

Elsberry, Wesley. 2007. *Logic and Math Turn to Smoke and Mirrors: William Dembski's Design Inference*. Scientists Confront Creationism: Intelligent Design and Beyond. Edited by Andrew J. Petto and Laurie R. Godfrey. New York: W. W. Norton.

Emergency Measure Number 1. 1974. Ministry of Public Administration and Security (body that maintains the Republic of Korea National Archive).

Emerson, Ralph Waldo. 1910. *Journals of Ralph Waldo Emerson: 1824-1832*. Edited by Edward Waldo Emerson and Waldo Emerson Forbes. London: Constable.

Emmerton, Ephraim. 1899. *Desiderius Erasmus of Rotterdam*. New York: The Knickerbocker Press.

Epictetus. 2003. *The Discourses of Epictetus*. First published circa 100 CE. Great Books of the Western World, Volume 11. Edited by Mortimer J. Adler, translated by George Long, seventh printing. Chicago: Encyclopædia Britannica.

Erskine, Thomas. 1792. *The Genuine Trial of Thomas Paine*. London: J. S Jordan.

Euripides. 2008. *Aegeus*. First published circa 460 BCE. Loeb Classical Library, LCL 504. Translated by Christopher Collard and Martin Cropp. Cambridge: Harvard University Press.

Everson v Board of Education. 1947a. United States Supreme Court. Justice Hugo Black leading, citation 330 US 1, 10 February 1947.

Everson v Board of Education. 1947b. United States Supreme Court. Justice Robert Jackson dissenting, citation 330 US 1, 10 February 1947.

Fahie, J. J. 1903. *Galileo: His Life and Work*. London: John Murray.

Fahringer, Catherine. 1991. Freedom From Religion. *San Antonio Express-News*. 24 March 1991.

Fallaci, Oriana. 1975. Indira's Coup. *The New York Review of Books*. 18 September 1975.

Farr, James, and Clayton Roberts. 1985. John Locke on the Glorious Revolution: A Rediscovered Document. *The Historical Journal*. Volume 28, number 2, June 1985, pp. 385-398.

Fecteau, Edward. 1945. *French Contributions to America*. Methuen: Soucy Press

Ferber, Edna. 1963. *A Kind of Magic*. Garden City: Doubleday.

Ferraù, Federico. 2012. *Scenario / Glendon (Harvard): l'Italia Impari dalla Crisi degli USA*. Il Sussidiario: Il Quotidiano Approfondito. 24 August 2012.

Festinger, Leon, et al. 2014. *When Prophecy Fails: A Social and Psychological Study of a Modern Group that Predicted the Destruction of the World*. Radford: Wilder Publications.

Feuerbach, Ludwig. 1848. *Das Wesen des Christenthums*. First published 1841. Leipzig: Otto Wigand.

Feynman, Richard. 1985. *Surely You're Joking, Mr. Feynman: Adventures of a Curious Character*. Edited by Richard Hutchings. New York: W. W. Norton.

Feynman, Richard. 1994. *The Character of Physical Law*. First published 1965. New York: Modern Library.

Fichte, Johann Gottlieb. 1806. *Die Grundzüge des Gegenwärtigen Zeitalters*. Berlin: Realschulbuchhandlung.

Fielding, Henry. 1902. *The Hearts of Men*. Second edition. London: Hurst and Blackett.

Finocchiaro, Maurice A. 1989. *The Galileo Affair: A Documentary History*.

Berkeley: University of California Press.
Fisher, Ronald. 1974. *The Design of Experiments*. First published 1935. Ninth edition. New York: MacMillan.
Flammarion, Camille. 1923. *Dreams of an Astronomer*. Translated by E. E. Fournier d'Albe. London: T. Fisher Unwin.
Foote, George William. 1893. *Flowers of Freethought*. London: R. Forder.
Forst de Battaglia, Otto, editor. 1970. *Dictatorship on Its Trial*. First published 1930. Translated by Huntley Paterson. Freeport: Books for Libraries Press.
Forster, E. M. 1979. *What I Believe*. First published 1938. Two Cheers for Democracy. New York: Harcourt Brace.
Fosdick, Harry Emerson. 1941. *Living Under Tension: Sermons on Christianity Today*. New York: Harper & Brothers.
Foucault, Michel. 2006. *History of Madness*. First published 1961. Translated by Jonathan Murphy and Jean Khalfa. London: Routledge.
Foxe, John. 1837. *The Acts and Monuments of John Foxe: A New and Complete Edition with a Preliminary Dissertation by the Rev. George Townsend. Volume 3*. First published 1570. London: Seeley and Burnside.
France, Anatole. 1895. *Le Jardin d'Épicure*. Paris: Calmann Lévy.
France, Anatole. 1906. *Le Lys Rouge*. First published 1894. Paris: Calmann Lévy.
France, Anatole. 1926. *Le Jardin d'Épicure*. First published 1894. Paris: Calmann Lévy.
Franklin, Benjamin. 1758. *Poor Richard Improved: Being an Almanack and Ephemeris of the Motions of the Sun and Moon*. July 1785, VII month. Philadelphia: B. Franklin and D. Hall.
Franklin's Motto. 1811. *The Monthly Anthology and Boston Review*. Volume 10, Silva 73, March 1811. Boston: T. B Wait.
Frazier, Kendrick. 1998. A Mind at Play: An Interview With Martin Gardner. *Skeptical Inquirer*. Volume 22, number 2, March/April 1998, pp. 34-39.
Freedom of Speech: Ghost at the Feast. 2019. *The Economist*. 31 August 2019.
Freeman, David. 2015. Neil deGrasse Tyson Sounds the Alarm Over Science Illiteracy, and for Good Reason. *Huffington Post*. 5 February 2015.
Freud, Sigmund. 1921. *Massenpsychologie und Ich-Analyse*. Leipzig: Internationaler Psychoanalytiker Verlag.
Freud, Sigmund. 1927. *Die Zukunft einer Illusion*. Leipzig: Internationaler

Psychoanalytiker Verlag.

Freud, Sigmund. 1930. *Das Unbehagen in der Kultur*. Vienna: Internationaler Psychoanalytiker Verlag.

Freud, Sigmund. 1933. *Neue Folge der Vorlesungen zur Einführung in die Psychoanalyse*. Vienna: Internationaler Psychoanalytiker Verlag.

Friedman, Milton, and Rose Friedman. 1980. *Free to Choose: A Personal Statement*. New York: Harcourt Brace Jovanovich.

Fromm, Erich. 1994. *Escape to Freedom*. First published 1941. New York: Henry Holt.

Fry, Stephen. 2001. *Room 101*. Interviewed by Paul Merton, season 6, episode 10, 12 March 2001. Aired by British Broadcasting Corporation (BBC). [video]

Fry, Stephen. 2004. *Paperweight*. First published 1992. London: Arrow.

Fulbright, J. William. 1966. *The Arrogance of Power*. New York: Random House.

Fuller, S. Margaret. 1845. *Woman in the Nineteenth Century*. New York: Greeley & McElrath.

Furet, François. 1999. *The Passing of an Illusion: The Idea of Communism in the Twentieth Century*. Translated by Deborah Furet. Chicago: University of Chicago Press.

Gage, Matilda Joslyn. 1893. *Woman, Church and State: A Historical Account of the Status of Woman Through the Christian Ages, with Reminiscences of Matriarchate*. Chicago: Charles Kerr.

Galilei, Galileo. 1623. *Il Saggiatore*. Rome: Appresso Giacomo Mascardi.

Galilei, Galileo. 1967. *Dialogue Concerning the Two Chief World Systems - Ptolemaic & Copernican*. Translated by Stillman Drake, second edition. Berkeley: University of California Press.

Gandhi, Mohandis. 1922. *Ethical Religion*. Translated by A. Rama Iyer, second edition. Madras: S. Ganesan.

Gandhi, Mohandis. 1924. *The Doctrine of the Sword*. Young India 1919-1922. Madras: S. Ganesan.

Gandhi, Mohandis. 1927. *Some Obiter Dicta: Some Posers*. Young India 1924-1926. Madras: S. Ganesan.

Gardner, Martin. 1996. *The Night Is Large: Collected Essays 1938-1995*. New York: St. Martin's Griffin.

Garrison, Wendell Phillips. 1885. *William Lloyd Garrison 1805-1879, The Story of His Life Told by His Children, Volume 1*. New York: Century.

Garrison, Fielding H. 1929. *An Introduction to the History of Medicine*. First published 1913. Fourth edition. Philadelphia: W. B Saunders.

Gawande, Atul. 2016. The Mistrust of Science. *The New Yorker*. 10 June 2016.

George, Alison. 2009. Eye-burrowing Worms, National Treasures … and Creationism. *New Scientist*. 28 January 2009. London: J. Johnson.

Gervais, Ricky. 2010. Why I Do Not Believe in a God. *The Wall Street Journal*. 19 December 2010.

Gervais, Ricky. 2013. *Twitter (rickygervais)*. 24 September 2013, 2:28am USA CST.

Ghosh, Prabir. 2008. *Why I Do Not Believe in God*. Translated by Sumitra Padmanabhan. Kolkata: Unique Ebook.

Gibbon, Edward. 2003. *The Decline and Fall of the Roman Empire, Volume 1*. First published 1776. Great Books of the Western World, Volume 37. Edited by Mortimer J. Adler, seventh printing. Chicago: Encyclopædia Britannica.

Gillispie, Charles Coulston, editor. 1980. *Dictionary of Scientific Biography, Volume 15, Supplement I*. New York: Charles Scribner's Sons.

Gleich, James. 1992. *Genius: The Life and Science of Richard Feynman*. New York: Vintage.

Goethe, Johann Wolfgang von. 1914. *Die Leiden des Jungen Werthers*. First published 1774. New York: Oxford University Press.

Goethe, Johann Wolfgang von. 1944. *Faust*. First published circa 1790. Goethe's Werke, Volume 3. Basel: Birkhäuser.

Goldman, Emma. 1913. The Failure of Christianity. *Mother Earth*. Volume 8, number 2, April 1913.

Goodall, Jane. 2002. The Power of One. *Time Magazine*. 26 August 2002.

Goodstein, Laurie. 2010. Basic Religion Test Stumps Many Americans. *The New York Times*. 28 September 2010.

Gorham, Charles T. 1930. *Religion as a Bar to Progress*. London: Watts.

Gould, Hialmer Day. 1905. *New Practical Spelling*. Cleveland: Practical Textbook Company.

Gould, Stephen Jay. 1992. *Bully for Brontosaurus*. First published 1992. New York: W. W. Norton.

Gould, Stephen Jay. 1995. *Adam's Navel and Other Essays*. London: Penguin.

Gould, Stephen Jay. 1997. *Full House: The Spread of Excellence from Plato to Darwin*. First published 1996. New York: Three Rivers Press.

Gould, Stephen Jay. 2007. *Wonderful Life: The Burgess Shale and the Nature of History*. First published 1989. New York: W. W. Norton.

Gourmont, Remy de. 1923. *Sixtine: Roman de la Vie Cérébrale*. First published 1890. Second edition. Paris: Mercure de France.

Grant, Ulysses S. 1870. *All Men Free and Equal: The Fifteenth Amendment Proclaimed - Message to Congress - Proclamation of the President*. Printed Ephemera, Three Centuries of Broadsides and Other Printed Ephemera (10,180). New Haven: J. H. Benham & Son, 30 March 1870. Washington, D.C.: Library of Congress.

Grayling, A. C. 2006. *The Heart of Things: Applying Philosophy to the 21st Century*. London: Orion.

Grayling, A. C. 2009. This Much I Know. *The Guardian*. 5 July 2009.

Green, Celia. 1977. *The Decline and Fall of Science*. First published 1976. Oxford: Institute of Psychophysical Research.

Green, Deven. 2016. Mississippi, the State Where It Violates Your Freedom to Be Made to Deliver a Wedding Cake. *Facebook*. Posted to photos wall as performance character Mrs. Betty Bowers, 5 April 2016.

Green, Deven. 2017. Happy Thanksgiving. *Facebook*. Posted to photos wall as performance character Mrs. Betty Bowers, 23 November 2017.

Gregory, T. Ryan. 2007. Orgel's Second Rule and "Unbeatable" Predation Tactics. *Genomicron*. 13 November 2007.

Grimmelshausen, Hans Jakob Christoffel von. 1899. *Der Abentheuerliche Simplicius Simplicissimus*. First published 1669. Grimmelshausen's Werke, Band 1, Erster Teil. Edited Felix Bobertag, published circa 1899, actual date not cited. Berlin: W. Spemann.

Guillen, Michael. 2004. *Can a Smart Person Believe in God?* Nashville: Nelson.

Gutnick, David. 2017. *Everybody Else's News: The Moose Jaw Times-Herald Closes Its Doors After 128 Years*. Canadian Broadcasting Corporation (CBC): The Sunday Edition. 1 December 2017.

The Hadith. First published circa 632 CE. [sunnah.com]

Haider, Sarah. 2019. *Twitter (sarahthehaider)*. 23 March 2019, 18:25 p.m. USA CST.

Haldane, J. B. S. 1934. *Fact and Faith*. London: Watts.

Haldane, J. B. S. 1963. The Truth About Death. *Journal of Genetics*. Volume 58, issue 3, December 1963.

Hallerberg, Arthur E. 1977. Indiana's Squared Circle. *Mathematics Magazine*. Volume 50, number 3, May 1977, pp. 136-140.

Hamilton, Alexander. 1904. *Speeches in the Federal Convention, 26 June 1787*. The Works of Alexander Hamilton, Volume 1. Edited by Henry Cabot Lodge. New York: G. P. Putnam's Sons.

Hamilton, Alexander, James Madison, and John Jay. 2003. *The Federalist*.

First published 1787-1788. Great Books of the Western World, Volume 40. Edited by Mortimer J. Adler, seventh printing. Chicago: Encyclopædia Britannica.

Harbottle, Thomas Harbottle. 1897. *Dictionary of Quotations (Classical)*. London: Swan Sonnenschein.

Hardy, G. H. 1967. *A Mathematician's Apology*. Cambridge: Cambridge University Press.

Harper, Ida Husted. 1898. *The Life and Work of Susan B. Anthony, Volume 2*. Indianapolis: Bowen-Merrill.

Harris, Sam. 2005a. *The End of Faith: Religion, Terror, and the Future of Reason*. New York: W. W. Norton.

Harris, Sam. 2005b. *The View from the End of the World*. Seminars About Long-Term Thinking (SALT). 9 December 2005. [video]

Harris, Sam. 2011. *The God Debate II*. Public debate held at the University of Notre Dame, Illinois, 7 April 2011. [video]

Hart, Bernard. 1914. *The Psychology of Insanity*. Cambridge: Cambridge University Press.

Hart, Jeffrey. 1966. *The American Dissent: A Decade of Modern Conservatism*. Garden City: Doubleday.

Hawking, Stephen. 1988. Gespräch mit dem Astrophysiker Stephen Hawking über Gott und das Weltall. *Der Spiegel*. Number 42, 17 October 1988.

Hawking, Stephen. 1993. *Black Holes and Baby Universes and Other Essays*. New York: Bantam.

Hawking, Stephen. 1998. *A Brief History of Time: Tenth Anniversary Edition*. First published 1988. New York: Bantam.

Hayek, Friedrich. 1944. *The Road to Serfdom*. London: George Routledge.

Hayek, Friedrich. 2011. *The Constitution of Liberty: The Definitive Edition*. First published 1960. The Collected Works of F. A. Hayek, Volume 17. Edited by Ronald Hamowy. Chicago: University of Chicago Press.

Hayes, Ellen. 1923. *How Do You Know? A Handbook of Evidence and Inference*. New York: Nation Press.

Hayman, Ronald. 2001. *A Life of Jung*. New York: W. W. Norton.

Hazlitt, William. 1902a. *Political Essays*. First published 1819. The Collected Works of William Hazlitt, Volume 3. Edited by A. R. Waller and Arnold Glover. London: J. M. Dent.

Hazlitt, William. 1902b. *The Round Table: On the Tendency of Sects*. First published 1815. The Collected Works of William Hazlitt, Volume 1. Edited by A. R. Waller and Arnold Glover. London: J. M. Dent.

Hazlitt, William. 1934. *Burke and the Edinburgh Phrenologists*. First published 1829. The Complete Works of William Hazlitt, Volume 20. Edited by P. P. Howe. London: J. M. Dent.

Heath, Thomas. 1913. *Aristarchus of Samos: The Ancient Copernicus*. Oxford: Clarendon Press.

Heehler, Tom. 2011. *The Well-Spoken Thesaurus: The Most Powerful Ways to Say Everyday Words and Phrases*. Naperville: Sourcebooks.

Heine, Heinrich. 1900. *Almansor*. First published 1882. Heinrich Heine: Sämtliche Werke in Zwölf Bänden, Fünfter Band. Estimated publication date of 1900, actual date not cited. Berlin: Theodor Knaur Nachfolger.

Heinlein, Robert. 1961. *Stranger in a Strange Land*. New York: Berkley.

Heinlein, Robert. 1966. *Life-Line*. The Worlds of Robert A. Heinlein. New York: Baen.

Heinlein, Robert. 1974. *Time Enough for Love: The Lives of Lazarus Long*. First published 1973. New York: Berkley Medallion.

Heinlein, Robert. 1986. *Revolt* in 2100. New York: Ace.

Hellman, Lillian. 1941. *Watch on the Rhein*. New York: Random House.

Hemingway, Ernest. 1995. *For Whom the Bell Tolls*. First published 1940. New York: Scribner.

Henry, Patrick. 1860. The Slave-Trade: An Original Letter from Patrick Henry. *The New York Times*. Volume 9, number 2745, 9 July 1860.

Herbert, George. 1841. *Jacula Prudentum*. First published 1640. The Remains of That Sweet Singer of the Temple. London: Pickering.

Herbert, Frank. 1976. *Children of Dune*. New York: Berkley Medallion.

Herbert, Frank. 1987. *God Emperor of Dune*. First published 1981. New York: Ace.

Herman, A. L. 1991. *A Brief Introduction to Hinduism*. Boulder: Westview Press.

Herodotus. 2003. *The Histories*. First published circa 440 BCE. Great Books of the Western World, Volume 5. Edited by Mortimer J. Adler, translated by George Rawlinson, seventh printing. Chicago: Encyclopædia Britannica.

Herzen, Aleksan. 1981. *From the Other Shore*. First published 1850. From the Other Shore, and The Russian People and Socialism: An Open Letter to Jules Michelet. Westport: Hyperion.

Hibben, John. 1908. The Paradox of Research. *The North American Review*. Volume 188, number 634, September 1908, pp. 425-431.

Hirsi Ali, Ayaan. 2006. *Hirsi Ali: Europe Has Double Standards on Israel*.

European Jewish Press. 5 September 2006.

Hirsi Ali, Ayaan. 2010. *Nomad: From Islam to America*. New York: Free Press.

Hitchens, Christopher. 1982. The Lord and the Intellectuals. *Harper's Magazine*. July 1982.

Hitchens, Christopher. 2007. *God is Not Great: How Religion Poisons Everything*. New York: Twelve / Hachette.

Hobbes, Thomas. 1953. *To My Most Honor'd Friend Mr. Francis Godolphin of Godolphin*. Leviathan. First published 1651. London: J. M. Dent.

Hobbes, Thomas. 2003. *Leviathan*. First published 1651. Great Books of the Western World, Volume 21. Edited by Mortimer J. Adler, seventh printing. Chicago: Encyclopædia Britannica.

Hobsbawm, Eric. 1994. *The Late Show*. British Broadcasting Corporation (BBC). Interview by Michael Ignatieff, 24 October 1994.

Hoffer, Eric. 1963. *The Ordeal of Change*. New York: Harper & Row.

Hoffer, Eric. 1974. *Reflections on the Human Condition*. New York: Harper & Row.

Hofstadter, Douglas. 1985. *Irrationality Is the Square Root of All Evil*. First published 1983. Metamagical Themas: Questing for the Essence of Mind and Pattern. New York: Basic.

Holmes, Oliver Wendell, Sr. 1891. *The Poet at the Breakfast Table, Volume 1*. First published 1872. Edinburgh: David Douglas.

Holmes, Oliver Wendell, Jr. 1919. Natural Law. *Harvard Law Review*. Volume 32, number 1, November 1918, pp. 40-44.

Holmes, John Haynes, et al., editors. 1919. *Readings From Great Authors*. New York: Dodd, Mead.

Holy, Ladislav. 1996. *The Little Czech and the Great Czech Nation: National Identity and the Post-Communist Social Transformation*. Cambridge: Cambridge University Press.

Hook, Sidney. 1940. *Reason, Social Myths and Democracy*. New York: John Day.

Hook, Sidney. 1950. Religion and the Intellectuals, Part II. *Partisan Review*. Volume 17, number 3, March 1950.

Hook, Sidney. 1964. *Religious Liberty from the Viewpoint of a Secular Humanist*. Religious Conflict in America: Studies of the Problems Beyond Bigotry. Edited by Earl Raab. Garden City: Anchor.

Horace. 1942. *Horace: Satires, Epistles and Ars Poetica*. First published circa 20 BCE. Loeb Classical Library, LCL 194. Translated by H. Rushton Fairclough. Cambridge: Harvard University Press.

Horace Mann League v Board of Public Works. 1966. Court of Appeals of Maryland. C. J. Prescott delivering, citation 242 Md. 645, 220 A.2d 51, 2 June 1966.

Horkheimer, Max. 1947. *Eclipse of Reason*. New York: Oxford University Press.

Horne, Charles Francis, editor. 1917. *On Medicine*. First published circa 1020. The Sacred Books and Early Literature of the East, with Historical Surveys of the Chief Writings of Each Nation, Volume 6. New York: Parke, Austin & Lipscomb.

Hossenfelder, Sabine. 2019. *Have we really measured gravitational waves?* 2 November 2019. [video]

Hubbard, Elbert. 1907. Heart to Heart Talks with Philistines by the Pastor of his Flock. *The Philistine*. Volume 25, number 3, August 1907.

Hubbard, Elbert. 1916. *The Philosophy of Elbert Hubbard*. East Aurora: Roycrofters.

Hubble, Edwin. 1954. *The Nature of Science and Other Lectures*. San Marino: Huntington Library.

Hugo, Victor. 1877. *Histoire d'un Crime: Déposition d'un Témoin*. Fourth edition. Paris: Ancienne Maison Michel Lévy Frères.

Hugo, Victor. 1878. *Discours de Victor Hugo*. Centenaire de Voltaire: Fête Oratoire. 30 May 1878. Paris: Bureaux du Siècle.

Human rights Lawyer Nasrin Sotoudeh Sentenced to 33 Years in Prison and 148 Lashes. 2019. *Front Line Defenders*. 21 March 2019.

Humboldt, Alexander von. 1845. *Kosmos: Entwurf Einer Physischen Weltbeschreibung - Volume 1*. Stuttgart: J. G. Cotta`sche.

Hume, David. 1767. *The Natural History of Religion*. First published 1757. Essays and Treatises on Several Subjects, Volume 2. London: A. Millar.

Hume, David. 1889. *Essays Moral, Political, and Literary, Volume 1*. First published 1758. Edited by T. H. Green and T. H. Grose. London: Longmans, Green.

Hume, David. 1912. *An Enquiry Concerning the Principles of Morals*. First published 1751. Chicago: Open Court.

Hume, David. 1960. *A Treatise of Human Nature*. First published 1740. Oxford: Clarendon Press.

Hume, David. 2003. *An Enquiry Concerning Human Understanding*. First published 1748. Great Books of the Western World, Volume 33. Edited by Mortimer J. Adler, seventh printing. Chicago: Encyclopædia Britannica.

Huxley, Thomas Henry. 1888. *Science And Culture*. London: MacMillan.

Huxley, Thomas Henry. 1898. *Biogenesis and Abiogenesis: The Presidential Address to the British Association for the Advancement of Science for 1870*. Discourses: Biological and Geological Essays, Volume 8. New York: D. Appleton.

Huxley, Thomas Henry. 1903. *On the Advisableness of Improving Natural Knowledge*. First published 1866. Lay Sermons, Addresses and Reviews. Third edition. New York: MacMillan.

Huxley, Thomas Henry. 1904. *Methods and Results*. First published 1893. Collected Essays, Volume 1. London: MacMillan.

Huxley, Julian. 1931. *Why I Reject the Idea of God*. Little Blue Book 1618. Edited by E. Haldeman-Julius. Girard: Haldeman-Julius.

Huxley, Aldous. 1957. *Proper Studies: The Proper Study of Mankind is Man*. First published 1927. London: Chatto & Windus.

Huxley, Aldous. 1958. *Brave New World Revisited*. New York: Harper & Row.

Huxley, Julian. 1961. *The Humanist Frame: The Modern Humanist Vision of Life*. New York: Harper & Brothers.

Ibsen, Henrik. 1999. An Enemy of the People. First published 1882. Mineola: Dover.

Iles, George. 1918. *Canadian Stories*. Montreal: Witness Press.

Ingersoll, Robert Green. 1894. *About the Holy Bible*. New York: C. P. Farrell.

Ingersoll, Robert Green. 1902a. *To the Indianapolis Clergy*. First published 1882. The Works of Robert G. Ingersoll, Volume 7. New York: Dresden.

Ingersoll, Robert Green. 1902b. *Why Am I an Agnostic?* First published 1896. The Works of Robert G. Ingersoll, Volume 11. New York: Dresden.

Ingersoll, Robert Green. 1909a. *Individuality*. First published 1878. The Works of Robert G. Ingersoll, Volume 1. New York: Dresden.

Ingersoll, Robert Green. 1909b. *Myth and Miracle*. The Works of Robert G. Ingersoll, Volume 2. New York: Dresden.

Ingersoll, Robert Green. 1909c. *What Must We Do to Be Saved?* First published 1880. The Works of Robert G. Ingersoll, Volume 1. New York: Dresden.

Ingersoll, Robert Green. 1921. *The Great Infidels: A Lecture*. New York: C. P. Farrell.

The Instructions of Shuruppag. 2001. The Electronic Text Corpus of Sumerian Literature. First published circa 2600 BCE. Author unknown. Oxford: University of Oxford.

An Intellectual Entente. 2009. *Harvard Magazine*. 10 September 2009.

Jacobs, Jane. 2004. *Dark Age Ahead*. New York: Vintage.

Jacobsen, Jacob. 2017. *Exclusive Interview with Aron Ra: Public Speaker, Atheist Vlogger, and Activist*. Conatus News. 13 May 2017.

Jacoby, Susan. 2018. *The Age of American Unreason in a Culture of Lies*. First published 2008. New York: Vintage.

James, William. 1902. *The Varieties of Religious Experience: A Study in Human Nature*. New York: Longmans, Green.

James, William. 1907. A Defense of Pragmatism. *The Popular Science Monthly*. Edited by J. McKeen Cattell, volume 70, March 1907. New York: Science Press.

James, William. 2003. *Pragmatism*. Great Books of the Western World, Volume 55. Edited by Mortimer J. Adler, seventh printing. Chicago: Encyclopædia Britannica.

Jameson, J. Franklin, editor. 1906. Papers of Dr. James McHenry on the Federal Convention of 1787. *The American Historical Review*. Volume 11, October 1905 to July 1906. New York: MacMillan.

Je suis Charlie. 2015. *Charlie Hebdo*. Number 1178, 14 January 2015.

Jefferson, Thomas. 1787. *Thomas Jefferson to David Hartley, 2 July 1787*. The Thomas Jefferson Papers. Series 1: General Correspondence 1651-1827. Washington, D.C.: Library of Congress.

Jefferson, Thomas. 1789. *Thomas Jefferson to Richard Price, 8 January 1789*. The Thomas Jefferson Papers. Series 1: General Correspondence 1651-1827. Washington, D.C.: Library of Congress.

Jefferson, Thomas. 1804. *Thomas Jefferson to John Tyler, 28 June 1804*. The Thomas Jefferson Papers. Series 1: General Correspondence 1651-1827. Washington, D.C.: Library of Congress.

Jefferson, Thomas. 1814. *Thomas Jefferson to Thomas Cooper, 7 October 1814*. The Thomas Jefferson Papers. Series 1: General Correspondence 1651-1827. Washington, D.C.: Library of Congress.

Jefferson, Thomas. 1816a. *Thomas Jefferson to Charles Yancey, 6 January 1816*. The Thomas Jefferson Papers. Series 1: General Correspondence 1651-1827. Washington, D.C.: Library of Congress.

Jefferson, Thomas. 1816b. *Thomas Jefferson to Francis Adrian Van der Kemp, 30 July 1816*. The Thomas Jefferson Papers. Series 1: General Correspondence 1651-1827. Washington, D.C.: Library of Congress.

Jefferson, Thomas. 1820a. *Thomas Jefferson to José Correa da Serra, 10 April 1820*. The Thomas Jefferson Papers. Series 1: General Correspondence 1651-1827. Washington, D.C.: Library of Congress.

Jefferson, Thomas. 1820b. *Thomas Jefferson to William Roscoe, 27 December 1820*. The Thomas Jefferson Papers. Series 1: General Correspondence 1651-1827. Washington, D.C.: Library of Congress.

Jefferson, Thomas. 1832. *Notes on the State of Virginia*. Boston: Lilly and Wait.

Jefferson, Thomas. 1903. *Notes on Virginia*. First published 1781. The Writings of Thomas Jefferson, Volume 2. Edited by Andrew A. Lipscomb and Albert Ellery Bergh, library edition. Washington, D.C.: Thomas Jefferson Memorial Association.

Jemison, Mae. 2005. *Then & Now: Dr. Mae Jemison*. CNN International. 19 June 2005.

Jobs, Steve. 2001. The Classroom of the Future. *Newsweek*. 28 October 2001.

Jones, LeRoi. 2009. *Home: Social Essays*. New York: Akashi Classics. [LeRoi Jones also known as Amiri Baraka]

Joyce, James. 1989. *Ireland, Island of Saints and Sages*. First published 1907. The Critical Writings of James Joyce. Ithaca: Cornell University Press.

Jung, Carl. 1970. *Alchemical Studies*. First published 1945. The Collected Works of C. G. Jung, Volume 13. Edited by Herbert Read, et al., translated by R. F C. Hull. Princeton: Princeton University Press.

Kaku, Michio. 1995. *Hyperspace: A Scientific Odyssey Through Parallel Universes, Time Warps, and the Tenth Dimension*. First published 1994. New York: Anchor.

Kaku, Michio. 2010. *Big Think Interview With Michio Kaku*. BigThink.com. Interviewed by Paul Hoffman, 11 October 2010. [audio transcript]

Kahneman, Daniel. 2011. *Thinking, Fast and Slow*. New York: Farrar, Strauss and Giroux.

Kanso, Heba. 2018. *Saudi Women Still Enslaved, Says Activist as Driving Ban Ends*. Thomson Reuters Foundation News. 22 June 2018.

Kant, Immanuel. 1784. Beantwortung der Frage: Was ist Aufklärung? *Berlinische Monatsschrift*. Volume 12, December 1784.

Kant, Immanuel. 1870. *Grundlegen zur Metaphysik der Sitten*. First published 1785. Berlin: L. Heimann.

Kant, Immanuel. 1920. *Prolegomena zu einer jeden künftigen Metaphysik, die als Wissenschaft wird auftreten können*. First published 1784. Edited by Karl Vorländer, sixth printing. Leipzig: Felix Meiner.

Kaufmann, Walter. 1963. *The Faith of a Heretic.* First published 1959. New York: Meridian.

Kaufmann, George, and Laurie Kauffman. 1994. Linus Pauling: Reflections. *American Scientist*. Volume 86, number 6, November-December 1994, pp. 522-524.

Kennedy, John F. 1960. *Speech to Ministerial Association, Houston, Texas, 12 September 1960*. Presidential Campaign Files 1960. John F. Kennedy Presidential Library and Museum.

Kennedy, John F. 1961. *Address at the University of Washington's 100th Anniversary Program*. Delivered 11 November 1961. John F. Kennedy Presidential Library and Museum.

Kennedy, John F. 1962a. *Address at Yale University, 11 June 1962*. John F. Kennedy Presidential Library and Museum.

Kennedy, John F. 1962b. *Remarks of the President at the Dinner Honoring All Living Nobel Prize Winners in the Western Hemisphere*. White House Staff Files of Sanford L. Fox,1961-1964, Nobel Prize Winner's Dinner, Folder 1 of 3, 29 April 1962. John F. Kennedy Presidential Library and Museum.

Kepler, Johannes. 1609. *Astronomica Nova*. Heidelberg: G. Voegelinus.

Ibn Khaldun. 1958. *The Muqaddimah: An Introduction to History, Volume I*. First published 1377. Translated by Franz Rosenthal. London: Routledge & Kegan Paul.

Kierkegaard, Søren. 1958. *De Omnibus Dubitandum Est*. Translated by T. H. Croxall. London: Adam & Charles Black.

King, Martin Luther, Jr. 1960. *The Rising Tide of Racial Consciousness, Address at the Golden Anniversary Conference of the National Urban League*. Delivered 6 September 1960, pp. 499–508. Stanford University Web Archive Portal: The Martin Luther King, Jr. Papers Project.

King, Martin Luther, Jr. 1963. *Letter from a Birmingham Jail*. Stanford University: The Martin Luther King, Jr. Research and Education Institute. 16 April 1963.

King, Martin Luther, Jr. 1965. *Speech at Brown Chapel in Selma, Alabama*. Delivered 8 March 1965. [video]

King, Stephen. 2012. *The Stand*. First published 1978. New York: Anchor.

King James Bible Online. First published 1611. [kingjamesbibleonline.org]

Kitzmiller v Dover Area School District. 2005. United States District Court, M.D. Pennsylvania. John Jones III presiding, citation 400 F. Supp. 2d 707, 20 December 2005.

Knight, Richard Payne. 1865. *On the Worship of Priapus*. First published 1786. Discourse on the Worship of Priapus and Its Connection With the Mystic Theology of the Ancients. London: Privately Printed.

Knight, Margaret K. 1955. *Morals Without Religion and Other Essays*. London: Dennis Dobson.

Knight, Damon. 1970. *Charles Fort: Prophet of the Unexplained*. Garden City: Doubleday.

Knight, Margaret K. 1995. *Christianity: The Debit Account*. First published 1975. London: National Secular Society.

Koestler, Arthur. 1943. A Challenge to Knights in Rusty Armor. *The New York Times*. 14 February 1943.

Koser, Reinhold, and Hans Droysen, editors. 1908a. *Briefwechsel Friedrichs des Grossen mit Voltaire, Erster Teil, Briefwechsel des Kronprinzen Friedrich 1736-1740*. Letter 4, Frederick to Voltaire, 4 November 1736. Leipzig: S. Hirzel.

Koser, Reinhold, and Hans Droysen, editors. 1908b. *Briefwechsel Friedrichs des Grossen mit Voltaire, Erster Teil, Briefwechsel des Kronprinzen Friedrich 1736-1740*. Letter 59, Frederick to Voltaire, June 1738. Leipzig: S. Hirzel.

Koupelis, Theo, and Karl F. Khun. 2007. *In Quest of the Universe*. First published 1991. Fifth edition. Sudbury: Jones and Bartlett.

Krauss, Lawrence. 2012. *A Universe From Nothing: Why There Is Something Rather Than Nothing*. New York: Free Press.

Krauthammer, Charles. 2018. A Note To Readers. *The Washington Post*. 8 June 2018.

Krucoff, Carol. 1984. The 6 O'Clock Scholar: Librarian of Congress Daniel Boorstin and His Love Affair With Books. *The Washington Post*. 29 January 1984.

Kurtz, Paul. 2012. *Exuberant Skepticism. For Good Reason*. Interviewed by D. J. Grothe, 3 February 2012. [audio podcast]

La Follette, Suzanne. 1926. *Concerning Women*. New York: Albert & Charles Boni.

Lactantius. 1543. *De Ira Dei*. First published circa 300 CE. L. Coelii Lactantii Firmiani Diuinarum Institutionum Libri Septem. London: Seb. Gryphium.

Laden, Osama bin. 2002. Letter to America. *The Guardian*. 24 November 2002.

Lamartine, Alphonse de. 1841. La Marseillaise de la Paix. *Revue des Deux-Mondes*. Volume 26, 1 June 1841, pp. 794-799.

Lamont, Corliss. 1957. *The Philosophy of Humanism*. First published 1949. Fourth edition. New York: Philosophical Library.

Landor, Walter Savage. 1891. *Imaginary Conversations, Volume 4*. First published 1829. London: J. M. Dent.

Lateran Treaty. 1929. The Holy See. Signed 11 February 1929, ratified 7 June 1929. [Trattato fra la Santa Sede e l'Italia]

Latham, Peter Mere. 1878. *The Collected Works of P. M. Latham, Volume 2*. Edited by Robert Martin. London: New Sydenham Society.

Laurie, Hugh. 2007. *The Right Stuff*. House. Voiced as character Gregory House, season 4, episode 2, 2 October 2007. Aired by National Broadcasting Company (NBC). [video]

Lavoisier, Antoine. 1793. *Traité Élémentaire de Chimie, Volume 1*. Second edition. Paris: Cuchet.

Lazarus, Emma. 1889. *The New Colossus*. The Poems of Emma Lazarus, Volume 1. Boston: Houghton Mifflin.

Lear, Linda, editor. 1998. *Lost Woods: The Discovered Writing of Rachel Carson*. Thorndike: Thorndike Press.

Lecky, W. E. H. 1886. *History of the Rise and Influence of the Spirit of Rationalism in Europe, Volume 2*. New York: D. Appleton.

Lecky, W. E. H. 1917. *History of European Morals: From Augustus to Charlemagne, Volume 1*. Third edition. New York: D. Appleton.

Lederman, Leon, and Dick Teresi. 1993. *The God Particle: If the Universe Is the Answer, What Is the Question?* Boston: Houghton Mifflin.

Lee, Robert E. 1856. *Robert E. Lee to Mary Anna Randolph, 27 December 1856*. Lee Family Digital Archive. Original letter at Virginia Museum of History & Culture, call number Mss1 L51 c 150-181.

Lehrer, Tom. 1965. *The Vatican Rag*. That Was The Year That Was (album). Reprise Records. [song lyric]

Lennon, John. 1971. *Imagine*. Imagine (album). Apple Records. [song lyric]

Lévi-Strauss, Claude. 1964. *Mythologiques: Le Cru et Le Cuit*. Paris: Plon.

Lewis, Sinclair. 1935. *It Can't Happen Here*. Garden City: Sun Dial Press.

Lewis, Joseph. 1938. *Shall Children Received Religious Instruction: A Debate Between Rev. Walter M. Howlett and Joseph Lewis*. First published 1932. Atheism and Other Addressess. New York: Freethought Press.

Lichtenberg, Georg Christoph. 1906. *Georg Christoph Lichtenbergs Aphorismen: Drittes Heft 1775-1779*. First published circa 1800. Deutsche Literaturdenkmale des 18. und 19. Jahrhunderts: No. 136, Dritte Folge No. 16. Edited by Albert Leitzmann. Berlin: B. Behr.

Lincoln, Abraham. 1864. *Abraham Lincoln to Albert G. Hodges, 4 April 1864*. The Abraham Lincoln Papers. Series 1. General

Correspondence. 1833-1916. Washington, D.C.: Library of Congress.

Lincoln, Abraham. 1898. *Gettysburg Address*. First published 1863. Abraham Lincoln: Early Speeches, Springfield Speech, Cooper Union Speech, Inaugural Addresses, Gettysburg Address, Selected Letters, Lincoln's Lost Speech. Edited by Bliss Perry, speech delivered 19 November 1863. New York: Doubleday & McClure.

Lindsey, Ben, and Wainwright Evans. 1935. *The Revolt Of Modern Youth*. First published 1928. London: John Lane.

Lippmann, Walter. 1915. *The Stakes of Diplomacy*. New York: Henry Holt.

Lippmann, Walter. 1944. *The Good Society*. First published 1937. Guildford: Billings.

Lippmann, Walter. 1964. *A Preface To Morals*. First published 1929. New York: Time-Life.

Locke, John. 1794. *A Letter to Anthony Collins, Esq*. The Works of John Locke, Volume 8. London: T. Longman & B. Law.

Locke, John. 1882. *Conduct of the Understanding*. First published 1706. Oxford: Clarendon Press.

Locke, John. 2003a. *Concerning Civil Government, Second Essay*. First published 1689. Great Books of the Western World, Volume 33. Edited by Mortimer J. Adler, seventh printing. Chicago: Encyclopædia Britannica.

Locke, John. 2003b. *An Essay Concerning Human Understanding*. First published 1689. Great Books of the Western World, Volume 33. Edited by Mortimer J. Adler, seventh printing. Chicago: Encyclopædia Britannica.

Locke, John. 2010. *An Essay Concerning Toleration*. A Letter Concerning Toleration and Other Writings. First published 1689. Edited by Mark Goldie. Indianapolis: Liberty Fund.

London, Scott. 2016. In the Footsteps of Gandhi: An Interview with Vandana Shiva. *The Ecologist: The Journal for the Post-Industrial Age*. 14 February 2016.

Lucas, J. R. 1979. Wilberforce and Huxley: A Legendary Encounter. *The Historical Journal*. Volume 22, number 2, June 1979, pp. 313-330.

Lucretius. 1910. *On the Nature of Things (De Rerum Natura)*. First published circa 50 BCE. Translated by Cyril Bailey. Oxford: Clarendon Press.

Luther, Martin. 1914. *Text der Predigten des Jahres 1546*. First published 1549. D. Martin Luthers Werke: Kritischer Gesamtausgabe, Volume 51. Sermon number 3, 17 January 1546. Weimar: Herman Böhlaus Nachfolger.

Luther, Martin. 1957. *Sermons on the Gospel of St. John, Chapters 1-4*. First published 1531-1537. Luther's Works, Volume 22. Chapter 3, sermon 26, edited by Jaroslav Pelikan. Saint Louis: Concordia.

Luther, Martin. 1959. *Sermons on the Gospel of St. John, Chapters 6-8*. First published 1531-1537. Luther's Works, Volume 23. Chapter 6, sermon 12, edited by Jaroslav Pelikan. Saint Louis: Concordia.

Luther, Martin. 1967. *The Christian in Society, Part 3: An Admonition to Peace, A Reply to the Twelve Articles of the Peasants in Swabia*. First published 1525. Luther's Works, Volume 46. Translated by Charles M. Jacobs, edited by Robert C. Schultz. Philadelphia: Fortress Press.

Luther, Martin. 1973. *Commentaries on 1 Corinthians 7, 1 Corinthians 15, Lectures on 1 Timothy*. First published 1531-1537. Luther's Works, Volume 28. Commentary on 1 Corinthians 7, edited by Milton Oswald. Saint Louis: Concordia.

Luxemburg, Rosa. 1922. *Die Russische Revolution: Eine Kritische Würdigung*. Berlin: Gesellschaft und Erziehung.

Macaulay, Thomas Babington. 1860. *Von Ranke*. Critical, Historical, and Miscellaneous Essays, Volume 4. New York: Sheldon. [First published in the *Edinburgh Review*, volume 72, number 145, October 1840, p. 227+]

Machiavelli, Niccolò. 1891. *Il Principe*. First published 1513. Edited by L. Arthur Burd. Oxford: Clarendon Press.

Macrobius. 1970. *Saturnalia*. Macrobius, Volume 1. Edited by J. Willis. Leipzig: B. G. Teubner.

Maddox, Brenda. 2002. *Rosalind Franklin: The Dark Lady of DNA*. New York: HarperCollins.

Madison, James. 1965. *Memorial and Remonstrance Against Religious Assessments*. First published 1785. Basic Documents Relating to the Religious Clauses of the First Amendment. Washington: Americans United for the Separation of Church and State.

Magnus, Albertus. 1967. *The Book of Minerals*. First published circa 1260. Translated by Dorothy Wyckoff. Oxford: Clarendon Press.

Maher, Bill. 2005. *I'm Swiss*. 30 July 2015. Aired by Home Box Office (HBO). [video]

Maistre, Joseph de. 1853. *Letter to A. M. Le Chevalier de X, 15 August 1811*. Lettres et Opuscules Inédits du Comte Joseph de Maistre, Volume 1. Second edition. Paris: A. Vaton.

Maistre, Joseph de. 1880. *Considérations sur la France*. First published 1796. Paris: Société Bibliographique.

Malala Yousafzai Rallies Youth to Stand Up for Universal Education. 2013. UN News. 12 July 2013.

Malcolm X. 1962. *Racial Militancy and Pride Urged at West Coast Rally*. Chicago Defender. 18 November 1962.

Mandela, Nelson. 1962. *Black Man in a White Court*. First court statement, 22 October 1962.

Mandela, Nelson. 2013. *Nelson Mandela by Himself: The Authorised Book of Quotations*. First published 2011. Johannesburg: MacMillan.

Manilius, Marcus. 1920. *Astronomicon, Book Four*. First published circa 10 CE. Commentary by A. E. Houseman. London: Grant Richards.

Mann, Thomas. 1967. *Der Zauberberg*. First published 1924. Frankfurt am Main: Fischer.

Mann, Thomas. 1969a. *An Appeal to Reason*. First published 1930. Order of the Day: Political Essays and Speeches of Two Decades. Translated by H. T. Lowe-Porter. Freeport: Books for Libraries Press.

Mann, Thomas. 1969b. *The War and the Future*. First published 1940. Order of the Day: Political Essays and Speeches of Two Decades. Translated by H. T. Lowe-Porter. Freeport: Books for Libraries Press.

Marlowe, Christopher. 1967. *The Jew of Malta*. First published circa 1589. Edited by T. W. Craik. New York: Hill and Wang.

Martineau, Harriet. 1837. *Society in America, Volume 3*. London: Saunders & Otley.

Marx, Karl. 2014. *Zur Kritik der Hegelschen Rechtsphilosophie*. First published 1814. Edited by Michael Holzinger. Berlin: CreateSpace.

Mason, George. 1776. *Virginia Declaration of Rights*. Commonwealth of Virginia. Ratified 12 June 1776.

Masood, Ehsan. 2009. *Science & Islam: A History*. London: Icon.

Massay, Gerald. 2019. *A Retort*. First published circa 1900. Gerald Massey's Lectures. Global Grey Ebooks.

Maugham, W. Somerset. 1915. *Of Human Bondage*. London: William Heinemann.

Maugham, W. Somerset. 1919. *The Moon and Sixpence*. New York: Grosset & Dunlap.

Maugham, W. Somerset. 1938. *The Summing Up*. New York: Literary Guild.

May, Thomas Erskin. 1878. Democracy in Europe: A History. *The Quarterly Review*. Volume 145, number 289, article 5, January & April 1878, pp. 112-142. London: John Murray.

May, Rollo. 1953. *Man's Search For Himself*. New York: Signet.

McCabe, McCabe. 1930. *Twelve Years in a Monastery*. London: Watts.

McCabe, Joseph. 1933. *The Existence of God*. London: Watts.
McCormack, Jerusha. 1997. *Wilde's Fictions*. The Cambridge Companion to Oscar Wilde. Edited by Peter Raby. Cambridge: Cambridge University Press.
McCreary County v American Civil Liberties Union of Kentucky. 2005. United States Supreme Court. Justice O'Connor concurring, citation 545 US 844, 27 June 2005.
McCurdy, Edward. 1923. *Leonardo Da Vinci's Note-Books: Arranged and Rendered Into English With Introductions*. New York: Empire State.
McGoohan, Patrick. 1967. *Arrival*. The Prisoner. Season 1, episode 1, 29 September 1967. Aired by Associated Television (ATV). [video - quote appears on blackboard in Labour Exchange office]
McKinney, Carlespie Mary Alice. 2010. *Why They Think I'm Crazy: Except When They Really Think About It*. Detroit: Folklore Entertainment.
McKown, Delos B. 1993. *The Mythmaker's Magic: Behind the Illusion of Creation Science*. Buffalo: Prometheus.
McLuhan, Marshall. 1977. *Marshall McLuhan in Conversation with Mike McManus*. Interview conducted on 19 September 1977. TVOntario. [video transcript]
Mead, Margaret. 1970. *Culture and Commitment*. New York: Natural History Press / Doubleday.
Medawar, P. B. 1981. *Advice To A Young Scientist*. First published 1979. New York: Basic.
Medawar, Peter. 1982. *Induction and Intuition in Scientific Thought*. First published 1969. Pluto's Republic. Oxford: Oxford University Press.
Mencken, H. L. 1908. *The Philosophy of Friedrich Nietzsche*. Boston: Luce.
Mencken, H. L. 1916. *A Little Book in C Major*. New York: John Lane.
Mencken, H. L. 1949. *A Mencken Chrestomathy*. New York: Alfred A. Knopf.
Mencken, H. L. 1956. *Minority Report: H. L. Mencken's Notebooks*. Baltimore: John Hopkins University Press.
Mencken, H. L. 2006. *On Politics: A Carnival of Buncombe*. First published 1956. Baltimore: John Hopkins University Press.
Merkel, Angela. 2009. Angela Merkels Rede im US-Kongress im Wortlaut. *Die Welt*. Speech delivered 4 November 2009.
Mettrie, Julien Offray de La. 1921. *L'Homme Machine*. First published 1748. Paris: Éditions Bossard.
Meyer, Holly. 2019. Harry Potter Books Removed From St. Edward Catholic School Due to Curses and Spells. *Tennessean*. 31 August 2019.

Mill, John Stuart. 1836. Civilisation. *The London and Westminster Review*. April-July 1836, pp. 1-28. London: William Clowes.

Mill, John Stuart. 1867. *Inaugural Address Delivered to the University of St. Andrews.* Delivered 1 February 1867. London: Longmans, Green, Reader, & Dyer.

Mill, John Stuart. 1874. *Theism*. Three Essays on Religion. New York: Henry Holt.

Mill, John Stuart. 2003. *On Liberty*. First published 1859. Great Books of the Western World, Volume 40. Edited by Mortimer J. Adler, seventh printing. Chicago: Encyclopædia Britannica.

Miller, Robin (Roblimo). 2004. *Wikipedia Founder Jimmy Wales Responds*. Slashdot. Posted 28 July 2004.

Miller, David. 2006. *Finding My Religion: Part I*. 10 April 2006. San Francisco Chronicle SFGate.

Mills, C. Wright. 1958. A Pagan Sermon to the Christian Clergy. *The Nation*. 8 March 1958, pp. 199-202.

Milton, John. 1911. *The Tenure of Kings and Magistrates*. First published 1649. New York: Henry Holt.

Mirzakhani, Maryam. 2014. The More I Spent Time on Maths, the More Excited I Got. *The Guardian*. 13 August 2014.

Mises, Ludwig von. 1947. *Planned Chaos*. Irvington-on-Hudson: Foundation for Economic Education.

Mises, Ludwig von. 1962. *The Ultimate Foundation of Economic Science: An Essay on Method*. Princeton: D. Van Nordstrand.

Mitchell, Edgar. 1979. *International Space Hall of Fame*. New Mexico Museum of Space History. Inducted 1979.

Mokyr, Joel. 2016. *A Culture of Growth: The Origins of the Modern Economy*. Princeton: Princeton University Press.

Molière. 1907. *Le Misanthrope*. First published 1666. Théatre de Molière, Volume 3. Paris: Bibliotèque National.

Montagu, Ashley, editor. 1984. *Science and Creationism*. Oxford: Oxford University Press.

Montaigne, Michel de. 1906. *Les Essais de Michel de Montaigne, Volume 3*. First published 1580. Bordeaux: Nouvelle F. Pech.

Montaigne, Michel de. 2003. *The Essays*. First published 1580. Great Books of the Western World, Volume 23. Edited by Mortimer J. Adler, seventh printing. Chicago: Encyclopædia Britannica.

Montesquieu. 1869. *Lettres Persanes*. Paris: Académie des Bibliophiles.

Montesquieu. 2003. *The Spirit of Laws*. First published 1784. Great Books of the Western World, Volume 35. Edited by Mortimer J. Adler, translated by J. V. Prichard, seventh printing. Chicago: Encyclopædia Britannica.

Montessori, Maria. 1912. *The Montessori Method*. Translated by Anne E. George. New York: Frederick A. Stokes.

Monty Python. 1979. *Life of Brian*. Voiced a character Brian Cohen. Python Pictures. [video]

Moore, Thomas. 1861. *Lalla Rookh: An Oriental Romance*. First published 1817. London: Longman, Green, Longman & Roberts.

More Than 1,000 People Detained During Opposition Election Protest in Moscow. 2019. Cable News Network (CNN). 27 July 2019.

Morgan, Augustus De. 1915. *A Budget of Paradoxes, Volume 2*. First published 1872. Edited by David Eugene Smith, second edition. Chicago: Open Court.

Morley, John. 1917. *On Compromise*. First published 1874. London: MacMillan.

Morris, Henry. 1978. *The Remarkable Birth of Planet Earth*. First published 1972. San Diego: Creation Life.

Morrison, Toni. 2012. *Peril*. First published 2009. Burn This Book: Notes on Literature and Engagement. New York: HarperCollins.

Motto. 2017. *The Washington Post*. First adopted February 2017.

Murasaki Shikibu. 2000. *The Tale of Genji*. First published circa 1000 CE. Translated by Arthur Waley. Mineola: Dover.

Murray, Gilbert. 1940. *Stoic, Christian and Humanist*. London: Watts.

Murray, Madalyn. 1965. Playboy Interview: Madalyn Murray. *Playboy*. October 1965.

Murrow, Edward R. 1967. *In Search of Light: The Broadcasts of Edward R. Murrow 1938-1961*. Edited by Edward Bliss. New York: Alfred A. Knopf.

The Nation's Best Bible College Gets Low Grades on Racial Diversity. 2001. *The Journal of Blacks in Higher Education*. Spring 2001, number 31, pp. 43-45.

New International Version Bible. First published 1978. [thenivbible.com]

New York Times v United States. 1971. United States Supreme Court. Justice Hugo Black leading, citation 403 US 713, 30 June 1971.

Newton, Isaac. 2003. *Quæstiones quædam Philosophiæ*. First published circa 1661. Trinity College Notebook. MS Add. 3996, Cambridge University Library, Cambridge, UK. [newtonproject.ox.ac.uk]

Niebuhr, Rheinhold. 1945. *The Children of Light and the Children of*

Darkness: A Vindication of Democracy and a Critique of Its Traditional Defenders. London: Nisbet.

Niebuhr, Rheinhold. 1953. *Christian Realism and Political Problems*. New York: Charles Scribner's Sons.

Niebuhr, Rheinhold. 2010. *Does Civilization Need Religion?* First published 1927. Eugene: Wipf and Stock.

Nietzsche, Friedrich. 1917. *Menschliches, Allzumenschliches, Erster Band*. First published 1878. Nietzsche's Werke, Volume 2. Leipzig: Alfred Kröner.

Nietzsche, Friedrich. 1919a. *Götzen-Dämmerung*. First published 1888. Nietzsche's Werke, Volume 8. Leipzig: Alfred Kröner.

Nietzsche, Friedrich. 1919b. *Umwerthung Aller Werthe*. First published 1888. Nietzsche's Werke, Volume 8. Leipzig: Alfred Kröner.

Nietzsche, Friedrich. 1921. *Menschliches, Allzumenschliches, Zweiter Band*. First published 1878. Nietzsche's Werke, Volume 3. Leipzig: Alfred Kröner.

Nietzsche, Friedrich. 1922a. *Der Will zur Macht: Erstes und Zweites Buch*. First published 1901. Nietzsche's Werke, Volume 15. Third edition. Leipzig: Alfred Kröner.

Nietzsche, Friedrich. 1922b. *Der Will zur Macht: Drittes und Viertes Buch*. Nietzsche's Werke, Volume 16. Second edition. Leipzig: Alfred Kröner.

Nine Inch Nails. 1994. *Heresy*. The Downward Spiral (album). Nothing Records. [song lyric]

Nisbet, Robert. 1982. *Prejudices: A Philosphical Dictionary*. Cambridge: Harvard University Press.

The Noble Qur'an. First published circa 632 CE. [quran.com]

North Carolina v Beal. 1930. Supreme Court of North Carolina. Justice Willam Stacy leading, citation 154 S.E. 604, 20 August 1930.

Norton, Charles Eliot. 1913. *Charles Eliot Norton to Goldwin Smith, 14 June 1897*. Letters of Charles Eliot Norton, Volume 2. Boston: Houghton Mifflin.

Notes and Clippings. 1889. *The Truth Seeker: A Journal of Freethought and Reform*. Volume 16, number 1, 5 January 1889.

Nye, Bill. 2012. *Creationism Is Not Appropriate For Children*. Big Think. 23 August 2012. [video]

Nye, Bill, and Ken Ham. 2014. *Is Creation A Viable Model of Origins?* Public debate held at Creation Museum, Kentucky, 4 February 2014. [video]

O'Connor, Flannery. 1983. *Everything That Rises Must Converge*. First published 1956. Three by Flannery O'Connor. New York: Signet.

O'Reilly, Jane. 1971. The Housewife's Moment of Truth. *New York Magazine*. 20 December 1971.

Obama, Barack. 2014. *Remarks by President Obama at Young Southeast Asian Leaders Initiative Town Hall*. The White House Press Office. 27 April 2014.

Ockham, William. 1995. *Prefatory Letter*. First published circa 1323. Summa Logicae. Translated Paul Vincent Spade.

Oliver, John. 2016. *Scientific Studies*. Last Week Tonight. Season 3, episode 11, 8 May 2016. Aired by Home Box Office (HBO). [video]

Olmstead v United States. 1928. United States Supreme Court. Justice Louis Brandeis dissenting, citation 277 US 438, 4 June 1928.

Onyanga-Omara, Jane. 2018. Stephen Hawking's Memorable Quotes. *USA Today*. 14 March 2018.

Origen of Alexandria. 1872. *The Writings of Origen, Volume 2: Origen Contra Celsum, Books II-VIII*. First published circa 248 CE. Ante-Nicene Christian Library, Translations of the Writings of the Fathers Down to A.D. 325, Volume 23. Translated by Frederick Crombie. Edinburgh: T. & T. Clark.

Orwell, George. 1968. *Letter to Humphry House*. The Collected Essays, Journalism and Letters of George Orwell: An Age Like This, Volume 1, 1920-1940. Edited by Sonia Orwell and Ian Angus. Boston: Nonpareil.

Orwell, George. 2003. *Nineteen Eighty-Four*. First published 1949. New York: Harcourt Brace.

Ouida. 1895. *The Failure of Christianity*. Views and Opinions. London: Methuen. [Ouida is the pen name of the writer Marie Louise de la Ramée]

Overstreet, Harry and Bonaro. 1964. *The Strange Tactics of Extremism*. New York: W. W. Norton.

Paine, Thomas. 1837. *Address to the Addressers on the Late Proclamation*. First published 1792. The Political Writings of Thomas Paine, Volume 2. Middletown: G. H. Evans.

Paine, Thomas. 1894. *The American Crisis*. First published 1776-1783. The Writings of Thomas Paine, Volume 1. Edited by Moncure Daniel Conway. New York: G. P. Putnam's Sons.

Paine, Thomas. 1896. *The Age of Reason*. First published 1794-1807. The Writings of Thomas Paine, Volume 4. Edited by Moncure Daniel Conway. New York: G. P. Putnam's Sons.

Paine, Thomas. 1908. *Worship and Church Bells*. First published 1797. The Life and Writings of Thomas Paine, Volume 7. Edited by Daniel Edwin

Wheeler. New York: Vincent Parke.

Paine, Thomas. 1935. *The Rights of Man*. First published 1791. London: J. M. Dent.

Palmer, Elihu. 1819. *Principles of Nature: Or a Development of the Moral Causes of Happiness and Misery Among the Human Species*. London: R. Carlile.

Pankhurst, Christabel. 1913. *The Great Scourge and How to End It*. London: E. Pankhurst.

Pankhurst, Emmeline. 1914. *My Own Story*. London: Eveleigh Nash.

Pareto, Vilfredo. 1919. *Manuale di Economia Politica con una Introduzione alla Scienza Sociale*. Milano: Societa Editrice Libraria.

Parfit, Derek. 1989. *Reasons and Persons*. First published 1984. Oxford: Clarendon Press.

Parker, Theodore. 1863. *Discourses on Politics*. The Collected Works of Theodore Parker, Volume 4. Edited by Frances Power Cobbe. London: Trübner.

Parks, Rosa. 1992. *Rosa Parks: My Story*. New York: Dial.

Parks, Rosa. 2005. *Civil Rights Icon Rosa Parks Dies*. National Public Radio (NPR). Main Reason for Keeping Her Seat, audio excerpt from 1992 interview with Lynn Neary, 25 October 2005.

Pascal, Blaise. 1854. *Sur l'Autorité en Matière de Philosophie, l'Esprit Géométrique et l'Art de Persuader*. First published 1658. Paris: L. Hachette.

Pascal, Blaise. 1918. *Pensées de Pascal, Volume 1*. First published 1670. Paris: Delagrave.

Pascal, Blaise. 2003. *Pensées*. First published 1670. Great Books of the Western World, Volume 30. Edited by Mortimer J. Adler, translated by W. F. Trotter, seventh printing. Chicago: Encyclopædia Britannica.

Passmore, John. 1970. *The Perfectibility of Man*. London: Duckworth.

Payne-Gaposchkin, Cecilia. 1984. *Cecilia Payne-Gaposchkin: An Autobiography and Other Recollections*. Edited by Katherine Haramundanis. Cambridge: Cambridge University Press.

Peirce, Charles Sanders. 1901. *The Century's Great Men in Science*. The 19th Century: A Review of Progress During the Past One Hundred Years in the Chief Departments of Human Activity. New York: G. P. Putnam's Sons.

Peirce, Charles Sanders. 1932. *Principles of Philosophy. Collected Papers of Charles Sanders Peirce, Volumes 1 and 2*. Edited by Charles Hartshorne and Paul Weiss. Cambridge: Belknap Press.

Penn, William. 1903. *Some Fruits of Solitude*. First published 1682. New York: Caldwell.

Pérez-Reverte, Artuno. 2006. *Purity of Blood*. Translated by Margaret Sayers Peden. London: Weidenfeld & Nicolson .

Peter, Laurence. 1977. *Peter's Quotations: Ideas for Our Time*. New York: William Morrow.

Petronius, Gaius. 2005. *Satyricon*. First published circa 50 CE. Loeb Classical Library, LCL 15. Translated by Michael Heseltine. Cambridge: Harvard University Press.

Pfeiffer, John. 1965. When Man First Stood Up. *The New York Times*. 11 April 1965.

Phillips, Wendell. 1852. *Speeches Before the Massachusetts Anti-Slavery Society*. Boston: Robert F. Wallcut.

Phillips, Wendell. 1884. *The Pilgrims*. First published 1855. Speeches, Lectures, and Letters. Speech delivered 21 December 1855. Boston: Lee and Shepard.

Pike, James A. 1964. *A Time for Christian Candor*. New York: Harper & Row.

Piketty, Thomas. 2014. *Capital in the Twenty-First Century*. Translated by Arthur Goldhammer. Cambridge: Belknap Press.

Pinker, Steven. 2009. *What is Your Dangerous Idea?* Edited by John Brockman. New York: HarperCollins.

Pinker, Steven. 2012. *The Better Angels of Our Nature: Why Violence Has Declined*. New York: Penguin.

Pinker, Steven. 2015. Why Free Speech Is Fundamental. *The Boston Globe*. 27 January 2015.

Pinto, Shiromi. 2017. *8 Women Show Us Why International Women's Day Is the Day to Declare: We Won't Wait for Our Rights!* Amnesty International. 5 March 2017.

Plait, Philip. 2015. *Introduction to Astronomy*. Astronomy Crash Course. Episode #1, 15 January 2015. Aired by PBS Digital Studios. [video]

Planck, Max. 1949. The Meaning and Limits of Exact Science. *Science*. Volume 110, issue 2857, 30 September 1949, pp. 319-327.

Plato. 2003a. *The Apology*. First published circa 380 BCE. Great Books of the Western World, Volume 6. Edited by Mortimer J. Adler, translated by Benjamin Jowett, seventh printing. Chicago: Encyclopædia Britannica.

Plato. 2003b. *The Republic*. First published circa 380 BCE. Great Books of the Western World, Volume 6. Edited by Mortimer J. Adler, translated by Benjamin Jowett, seventh printing. Chicago: Encyclopædia Britannica.

Plato. 2003c. *Theaetetus*. First published circa 370 BCE. Great Books of the Western World, Volume 6. Edited by Mortimer J. Adler, translated by Benjamin Jowett, seventh printing. Chicago: Encyclopædia Britannica.

Plato. 2003d. *Protagoras*. First published circa 370 BCE. Great Books of the Western World, Volume 6. Edited by Mortimer J. Adler, translated by Benjamin Jowett, seventh printing. Chicago: Encyclopædia Britannica.

Poincaré, Henri. 1917. *La Science et l'Hypothèse*. First published 1902. Paris: Ernest Flammarion.

Poincaré, Henri. 1920. *Science et Methode*. Paris: Ernest Flammarion.

Pope John Paul II. 1996. *Mensajes del Santo Padre Juan Pablo II a los Miembros de la Academia Pontificia de Ciencias*. The Holy See. 22 October 1996.

Pope Leo XIII. 1888. *Libertas: Encyclical of Pope Leo XIII on the Nature of Human Liberty*. The Holy See. 20 June 1888.

Pope Leo XIII. 1901. *Graves de Communi Re: Encyclical of Pope Leo XIII on Christian Democracy*. The Holy See. 18 January 1901.

Pope Paul VI. 1964. *Lumen Gentium: Dogmatic Constitution on the Church*. The Holy See. 21 November 1964.

Popper, Karl. 1947. *The Open Society and Its Enemies, Volume 1*. First published 1945. London: Routledge Classics.

Popper, Karl. 1965. *Truth, Rationality, and the Growth of Scientific Knowledge*. Conjectures and Refutations: The Growth of Scientific Knowledge. First published 1963. Second edition. New York: Basic.

Popper, Karl. 2002. *The Logic of Scientific Discovery*. First published 1959. London: Routledge Classics.

Pratchett, Terry. 2003. *Monstrous Regiment: A Novel of Discworld*. New York: HarperCollins.

Priestly, J. B. 1947. *The Arts Under Socialism*. London: Turnstile Press.

Prince v Massachusetts. 1944. United States Supreme Court. Justice Rutledge leading, citation 321 US 158, 31 January 1944.

Province of Massachusetts. 1742. *The Charter Granted by their Majesties King William and Queen Mary, to the Inhabitants of the Province of the Massachusetts-Bay in New-England*. New England: Bolton.

Pullman, Phillip. 2002. *A Dark Agenda?* Surefish Interview. November 2002.

Qin Yongmin: Prominent Chinese Dissident Jailed for 13 Years. 2018. *BBC*

News. 11 July 2018.

Ra, Aron. 2012. *Don't Blame the Atheists*. Texas Freethought Convention. Speech delivered 20 October 2012.

Ra, Aron. 2015a. *Religious Indoctrination Rampant in Rural Texas Schools*. Patheos. 28 October 2015.

Ra, Aron. 2015b. *Anti-Theist Answers to Christian Questions*. Patheos. 27 November 2015.

Ra, Aron. 2017. *Philosophistry*. Patheos. 12 April 2017.

Rabelais, François. 1913. *Gargantua et Pantagruel, Volume 1*. First published circa 1532. Paris: Larousse.

Rabi, Isidor Isaac. 1964. Humanistic Scientist. *The New York Times*. 28 October 1964.

Ramasamy, V. R. 2007. *Collected Works of Periyar E. V. R.* Chennai: The Periyar Self-Respect Propaganda Institution.

Rand, Ayn. 1967. *Capitalism: The Unknown Ideal*. New York: Signet.

Rand, Ayn. 1990. *Epitaph for a Culture*. The Voice of Reason: Essays in Objectivist Thought. Edited by Leonard Peikoff. New York: Meridian.

Rao, Goparaju Ramachandra. 1978. *Positive Atheism*. First published 1972. Vijayawada: Atheist Centre.

Ratner, Sidney. 1936. Evolution and the Rise of the Scientific Spirit in America. *Philosophy of Science*. Volume 3, number 1, January 1936, pp. 104-122

Rauf, Feisal Abdul. 2012. *Moving The Mountain: A New Vision of Islam in America*. New York: Free Press.

Rawls, John. 1971. *A Theory of Justice*. Cambridge: Belknap Press.

Reade, Winwood. 1887. *The Martyrdom of Man*. First published 1872. Twelfth edition. London: Trübner.

Reagan, Ronald. 1987. *Remarks at the Annual Convention of Kiwanis International*. Ronald Reagan Presidential Foundation & Institute. Speech delivered 6 July 1987.

Reichskonkordat. 1933. Signed on 20 July 1933 by Eugenio Pacelli, who later became Pope Pius XII, on behalf of Pope Pius XI.

Reeves, Keanu. 2019. *The Late Show with Stephen Colbert*. Season 4, episode 145, 10 May 2019. Aired by Columbia Broadcasting System (CBS). [video]

Renan, Ernest. 1863. *Vie de Jésus*. Paris: Michel Lévy.

Rich, Adrienne. 1975. *Snapshots of a Daughter-in-Law*. First published 1963. Adrienne Rich's Poetry: Texts of the Poems, The Poet on Her Work, Reviews and Criticism. Edited by Barbara Charlesworth Gelpi and Albert Gelpi. New York: Norton.

Rittner, Carol, and Sondra Myers, editors. 1986. *The Courage to Care: Rescuers of Jews During the Holocaust*. New York: New York University Press.

Robertson, John M. 1906. *A Short History of Freethought: Ancient and Modern, Volume 1*. Second edition. New York: G. P. Putnam's Sons.

Robertson, James. 2018. *Debate Over Confederate Monuments*. Speech delivered at Shenandoah Valley Battlefields Foundation conference on 28 July 2018. Aired by C-SPAN3. [video]

Rochefoucauld, François de La. 1923. *Réflexions ou Sentences et Maximes Morales*. First published circa 1670. Œuvres de La Rochefoucauld, Volume 1. Paris: Hachette.

Roddenberry, Gene. 1967a. *The Squire of Gothos*. Star Trek. Performance by Leonard Nimoy, season 1, episode 17, 12 January 1967. Aired by National Broadcasting Company (NBC). [video]

Roddenberry, Gene. 1967b. *Arena*. Star Trek. Performance by Carole Shelyne, season 1, episode 18, 19 January 1967. Aired by National Broadcasting Company (NBC). [video]

Roosevelt, Theodore. 1908. *Letter to J. C. Martin. 6 November 1908*. Almanac of Theodore Roosevelt. [theodore-roosevelt.com]

Roosevelt, Eleanor. 1935. *In Defense of Curiosity*. The Saturday Evening Post. 24 August 1935.

Roosevelt, Franklin. 1938. *Address at Temple University - Honorary Degree*. The Public Papers and Addresses of Franklin D. Roosevelt, Volume 5. 22 February 1936. New York: Random House.

Roosevelt, Eleanor. 1949. *May Day*. The Eleanor Roosevelt Papers Digital Edition. 8 July 1949. The George Washington University.

Root-Bernstein, Robert S. 1995. Darwin's Rib. *Discover*. September 1995, pp. 38-41.

Rosenfeld, Albert. 1966. *The Quintessence of Irving Langmuir*. Selected Readings in Physics: Men of Physics. Oxford: Pergamon.

Rosenfeld, Megan. 1987. Appeal of the Televangelists: Firm Answers to Life's Questions. *The Washington Post*. 3 May 1987.

Rousseau, Jean-Jacques. 1912. *Émile, Volume 1*. First published 1762. Paris: Jean Gillequin.

Rousseau, Jean-Jacques. 2003. *The Social Contract*. First published 1762. Great Books of the Western World, Volume 35. Edited by Mortimer J. Adler, translated by G. D. H. Cole, seventh printing. Chicago: Encyclopædia Britannica.

Roy, Manabendra Nath. 1981. *Practice of New Humanism*. First published 1948. Statement adopted by the All-India Conference of the Radical Democratic Party held in Calcutta, 26-30 December 1948. New Dehli: Hindustan Offset Press.

The Royal Society. [royalsociety.org]

Royce, Josiah. 1934. *Lectures on Modern Idealism*. First published 1919. Third printing. New Haven: Yale University Press.

Rubin, Dave. 2015. *Dave Rubin Explains The Rubin Report Rules*. The Rubin Report. 9 September 2015. [video]

Ruse, Michael. 1982. *Darwinism Defended: A Guide to the Evolution Controversies*. Reading: Addison-Wesley.

Rushdie, Salman. 1989. *The Satanic Verses*. New York: Viking.

Rushdie, Salman. 1990. *Is Nothing Sacred?* Herbert Reade Memorial Lecture, 6 February 1990. London: Granta.

Ruskin, John. 1890. T*he Stones of Venice: Volume 2, The Sea Stories*. Boston: Aldine.

Ruskin, John. 1909. *The Letters of John Ruskin, Volume 1, 1827-1869*. The Works of John Ruskin, Volume 36. Edited by E. T. Cook and Alexander Wedderburn, letter from Ruskin to Henry Acland, 24 May 1851. London: George Allen.

Russell, Bertrand. 1930. *Marriage and Morals*. London: George Allen & Unwin.

Russell, Bertrand. 1945. *A History of Western Philosophy*. New York: Simon & Schuster.

Russell, Bertrand. 1950. *An Outline of Intellectual Rubbish*. Unpopular Essays. First published 1921. New York: Simon & Schuster.

Russell, Bertrand. 1953. *The Impact of Science on Society*. New York: Simon & Schuster.

Russell, Bertrand. 1957. *Why I Am Not a Christian: And Other Essays on Religion and Related Subjects*. New York: Simon & Schuster.

Russell, Bertrand. 1958. *Voltaire's Influence on Me*. Studies on Voltaire and the Eighteenth Century, Volume 6. Edited by Theodore Besterman. Genève: Institut et Musée Voltaire.

Russell, Bertrand. 1959. *Mysticism and Logic and Other Essays*. First published 1910. London: George Allen & Unwin.

Russell, Bertrand. 1998. *The Triumph of Stupidity*. First published 1933. Mortals and Others, Volume 2: American Essays, 1931-1935. London: Routledge.

Rutherford, Adam. 2017. *A Brief History of Everyone Who Ever Lived: The*

Human Story Retold Through Our Genes. New York: The Experiment.

Ryan, Edward. 1943. Address to the Graduating Law Students of the University of Wisconsin, 1873. *Notre Dame Law Review*. Volume 19, issue 2, 1 December 1943.

Safire, William. 1992. *Lend Me Your Ears: Great Speeches in History*. New York: W. W. Norton.

Sagan, Carl. 1979. *Broca's Brain: Reflections on the Romance of Science*. New York: Ballantine.

Sagan, Carl. 1980. *Cosmos*. First published 1980. New York: Random House.

Sagan, Carl. 1987. The Burden of Skepticism. *Skeptical Inquirer*. Volume 12, number 1, Fall 1987.

Sagan, Carl. 1990. Why We Need to Understand Science. *Skeptical Inquirer*. Volume 14, issue 3.

Sagan, Carl. 1994a. *Pale Blue Dot: A Vision of the Human Future in Space*. New York: Ballantine.

Sagan, Carl. 1994b. With Science On Our Side. *The Washington Post*. 9 January 1994.

Sagan, Carl. 1996. *The Demon-Haunted World: Science as a Candle in the Dark*. New York: Ballantine.

Sagan, Carl. 2000. *Encyclopedia Galactica*. Cosmos: A Personal Voyage. Episode 12. Aired by Public Broadcasting Service (PBS). [video]

St. John, Robert. 1972. *Eban*. Garden City: Doubleday.

Sakharov, Andrei. 1968. *Progress, Coexistence, and Intellectual Freedom*. Translated by The New York Times. New York: W. W. Norton.

Salt, Henry Stephens. 1921. *Seventy Years Among Savages*. London: George Allen & Unwin.

Sand, George. 1873. *Impressions et Souvenirs*. Qeuvres de George Sand, Volume 93. Paris: Michel Lévy.

Sanger, Margaret. 1914. *The Woman Rebel*. Newsletter motto, volume 1, number 1, March 1914.

Sanger, Margaret. 1921. *Birth Control: What It Is, How It Works, What It Will Do*. Proceedings of the first American birth control conference held at the Hotel Plaza, New York, 11-12 November 1921. New York: Birth Control Review.

Santayana, George. 1922. *On My Friendly Critics*. Soliloquies in England and Later Soliloquies. London: Constable.

Santayana, George. 1929. *Scepticism and Animal Faith: Introduction to a System of Philosophy*. First published 1923. New York: Charles

Scribner's Sons.

Santayana, George. 1931a. *Imaginative Nature of Religion*. Little Essays Drawn From the Writings of George Santayana. First published 1920. New York: Charles Scribner's Sons.

Santayana, George. 1931b. *Christian Morality*. Little Essays Drawn From the Writings of George Santayana. First published 1920. New York: Charles Scribner's Sons.

Sarton, George. 1988. *The History of Science and the New Humanism*. First published 1962. New Brunswick: Transaction Books.

Sartre, Jean-Paul. 1965. *Anti-Semite and Jew*. First published 1946. New York: Schocken.

Schenck v United States. 1919. United States Supreme Court. Justice Oliver Wendell Holmes leading, citation 249 US 47, 3 March 1919.

Schick, Ludwig. 2012. Bischof Fordert Anti-Blasphemie-Gesetz. *Süddeutsche Zeitung*. 2 August 2012.

Schiller, Friedrich von. 1878. *Geschichte des Abfalls der Vereinigten Niederlande von der Spanischen Regierung*. First published 1788. Stuttgart: J. G. Cotta`sche.

Schlegel, Friedrich. 1904. *Fragmente*. Edited by Friedrich von der Leyen. Leipzig: Eugen Dietrichs.

Schlesinger, Arthur, Jr. 1989. The Opening of the American Mind. *The New York Times*. 23 July 1989.

Schnitlzer, Arhtur. 1967. *Buch der Sprüche und Bedenken*. First published circa 1925. Aphorism und Betrachtungen. Frankfurt am Main: Fischer.

Schopenhauer, Arthur. 1920a. *Die Welt als Wille und Vorstellung: Ergänzungen zum Vierten Buch*. First published 1844. Schopenhauer's Sämtliche Werke in Fünf Bänden, Volume 2. Grossherzog Wilhelm Ernst edition. Leipzig: Insel.

Schopenhauer, Arthur. 1920b. *Parerga und Paralipomena: Über das Christenthum*. First published 1851. Schopenhauer's Sämtliche Werke in Fünf Bänden, Volume 5. Grossherzog Wilhelm Ernst edition. Leipzig: Insel.

Schopenhauer, Arthur. 1920c. *Parerga und Paralipomena: Über Religion*. First published 1851. Schopenhauer's Sämtliche Werke in Fünf Bänden, Volume 5. Grossherzog Wilhelm Ernst edition. Leipzig: Insel.

Schurz, Carl. 1913a. *The Policy of Imperialism*. Speeches, correspondence and political papers of Carl Schurz, Volume 6. Speech delivered 17 October 1899, edited by Frederic Bancroft. New York: G. P. Putnam's Sons.

Schurz, Carl. 1913b. *The Venezuela Question*. Speeches, correspondence and political papers of Carl Schurz, Volume 5. Speech delivered 2 January 1896, edited by Frederic Bancroft. New York: G. P. Putnam's Sons.

Schwarz, Maralee, and Kenneth J. Cooper. 1992. Equal Rights Initiative in Iowa Attacked. *Washington Post*. 23 August 1992.

Seaver, George. 1949. Albert Schweitzer: *The Mind and His Mind*. Third edition. London: Adam & Charles Black.

Sen, Amartya. 2006. *Identify and Violence: The Illusion of Destiny*. New York: W. W. Norton.

Seneca The Younger. 1953. *Letters From a Stoic (Epistulae Morales ad Lucilium)*. First published circa 60 CE. Loeb Classical Library, LCL 77. Translated by Richard M. Gummere. Cambridge: Harvard University Press.

Seneca The Younger. 1985. *On Anger (De Ira)*. First published circa 40 CE. Loeb Classical Library, LCL 214. Translated by John W. Basore. Cambridge: Harvard University Press.

Seneca The Younger. 2002. *Medea*. First published circa 50 CE. Loeb Classical Library, LCL 62. Translated by John G. Fitch. Cambridge: Harvard University Press.

Servetus, Michael. 1969. *On the Righteousness of Christ's Kingdom*. First published circa 1532. The Two Treatises of Servetus on the Trinity. First published 1932. Translated by Earl Morse Wilbur. New York: Krauss.

Sextus Empiricus. 1933. *Outlines of Pyrrhonism*. First published circa 200 CE. Loeb Classical Library, LCL 273. Translated by R. G. Bury. Cambridge: Harvard University Press.

Shakespeare, William. 2003a. *A Comedy of Errors*. First published 1623. Great Books of the Western World, Volume 24. Edited by Mortimer J. Adler, seventh printing. Chicago: Encyclopædia Britannica.

Shakespeare, William. 2003b. *Julius Caesar*. First published 1599. Great Books of the Western World, Volume 24. Edited by Mortimer J. Adler, seventh printing. Chicago: Encyclopædia Britannica.

Shaw, George Bernard. 1919. *Heartbreak House, Great Catherine, and Playlets of the War*. London: Constable.

Shaw, George Bernard. 1921. *Back to Methuselah: A Metabiological Pentateuch*. New York: Brentano's.

Shaw, George Bernard. 1957. *Androcles and the Lion*. First published 1913. London: Penguin.

Shaw, George Bernard. 1958. *The Admirable Bashville*. First published 1901. Seven One-Act Plays. Baltimore: Penguin.

Sheed, Wilfrid. 1985. *Frank and Maisie: A Memoir With Parents*. New York: Simon & Schuster.

Shelley, Percy Bysshe. 1904. *Notes to Queen Mab*. First published 1813. The Complete Works of Percy Bysshe Shelley, Volume 1. London: Virtue.

Shelley, Percy Bysshe. 1915a. *Letter to Lord Ellenborough*. First published 1812. Selected Prose: Works of Shelley. London: Watts.

Shelley, Percy Bysshe. 1915b. *A Refutation of Deism*. First published 1814. Selected Prose: Works of Shelley. London: Watts.

Shermer, Michael. 2019. *Twitter (michaelshermer)*. 30 January 2019, 5:42am USA CST.

Shlachter, Barry. 1985. Woman Fierce Protector of Gorilla Friends. *Lawrence Journal-World*. 29 July 1985.

Sim, Stuart. 2006. *Empires of Belief: Why We Need More Scepticism and Doubt in the Twenty-First Century*. Edinburgh: Edinburgh University Press.

Sime, James. 1879. *Lessing, Volume 2*. First published 1877. Second edition. London: Trübner.

Sime, Ruth Lewin. 1996. *Lise Meitner: A Life in Physics*. Berkeley: University of California Press.

Simpson, George Gaylord. 1964. *This View of Life: The World of an Evolutionist*. New York: Harcourt, Brace & World

Sinclair, Upton. 1918. *The Profits of Religion: An Essay in Economic Interpretation*. First published 1917. Pasadena: Self-Published.

Singer, Peter. 1981. *The Expanding Circle: Ethics and Sociobiology*. New York: Meridian.

Singh, Simon. 2005. *Big Bang*. Hammersmith: Harper Perennial.

Smith, Preserved, and Charles M. Jacobs. 1918. *The Estates of the Empire to Francis Chieregato, 5 February 1523*. Luther's Correspondence and Other Contemporary Letters, Volume 2, 1521-1530. Philadelphia: Lutheran Publication Society.

Smith, Margaret Chase. 1952. We Must Not Forfeit Freedom. *NEA Journal*. Volume 41, number 5, May 1952 (from speech delivered at Hood College on 21 April 1951.

Smith, Preserved. 1957. *A History of Modern Culture, Volume 1, The Great Renewal, 1543-1687*. Gloucester: Peter Smith.

Smith, Margaret Chase. 1972. *Declaration of Conscience*. Speech delivered

1 June 1950, edited by William C. Lewis, Jr. Garden City: Doubleday.
Smith, George. 1973. *On Self Discovery and Self Responsibility: An Interview With Nathaniel Branden*. May 1973. [reason.com]
Smith, John Maynard. 1976. Group Selection. *The Quarterly Review of Biology*. Volume 51, number 2, June 1976, pp 277-283.
Smith, George H. 1991. *Atheism, Ayn Rand, and Other Heresies*. New York: Prometheus.
Smith, Lilian. 1994. *Killers of the Dream*. First published 1949. New York: W. W. Norton.
Smith, Adam. 2003. *An Inquiry Into the Nature and Causes of the Wealth of Nations*. First published 1776. Great Books of the Western World, Volume 36. Edited by Mortimer J. Adler, seventh printing. Chicago: Encyclopædia Britannica.
Snyder, Timothy. 2017. *On Tyranny: Twenty Lessons from the Twentieth Century*. New York: Penguin Random House.
Solzhenitsyn, Aleksandr. 1974. Solzhenitsyn: An Artist Becomes an Exile. *Time*. Volume 103, number 8, 25 February 1974.
Soothill, William Edward. 1910. *The Analects of Confucius*. First published 1677. Yokohama: Fukuin.
Sowell, Thomas. 1996. *Knowledge and Decisions*. First published 1980. New York: Basic.
Spencer, Herbert. 1883. *Social Statics: Or the Conditions Essential to Human Happiness Specified and the First of Them Developed*. New York: D. Appleton.
Spencer, Herbert. 1898. *The Principles of Ethics, Volume 1*. New York: D. Appleton.
Spencer, Herbert. 1907. *The Development Hypothesis*. First published 1852. Seven Essays Selected from the Works of Herbert Spencer. London: Watts.
Spinelli, Altiero, and Ernesto Rossi. 2017. *Il Manifesto di Ventotene*. First published 1941. Torino: Celid.
Spinoza, Benedict de. 1862. *Tractatus Theologico-Politicus: A Critical Inquiry into the History, Purpose, and Authenticity of the Hebrew Scriptures; with the Right to Free Thought and Free Discussion*. First published 1670. London: Trübner.
Spinoza, Benedict de. 2003. *Ethics*. First published 1677. Great Books of the Western World, Volume 28. Edited by Mortimer J. Adler, seventh printing. Chicago: Encyclopædia Britannica.
Stanislaw, Lec. 1962. *Unkempt Thoughts*. Translated by Jacek Galazka.

New York: St. Martin's Press.

Stanislaw, Lec. 1968. *More Unkempt Thoughts*. Translated by Jacek Galazka. New York: Funk & Wagnalls.

Stanley v Georgia. 1969. United States Supreme Court. Justice Thurgood Marshall leading, citation 394 US 557, 7 April 1969.

Stanton, Elizabeth Cady, et al. 1848. *Report of the Woman's Rights Convention, held at Seneca Falls, New York, July 19th and 20th, 1848: Proceedings and Declaration of Sentiments*. Miller NAWSA Suffrage Scrapbooks, 1897-1911, Scrapbook 6 (1907-1908). Washington, D.C.: Library of Congress.

Stanton, Elizabeth Cady, et al., editors. 1887a. *History of Woman Suffrage, Volume 1, 1848-1861*. Rochester: Charles Mann.

Stanton, Elizabeth Cady, et al., editors. 1887b. *History of Woman Suffrage, Volume 3, 1876-1885*. Rochester: Charles Mann.

Stanton, Elizabeth Cady. 1993. *The Woman's Bible*. First published 1895. Boston: Northeastern University Press.

Steinem, Gloria. 1986. *Outrageous Acts and Everyday Rebellions*. First published 1983. New York: Signet.

Steinmetz, Charles. 1922. Religion and Modern Science: How They Differ, and the Place of Each. *Christian Register*. Volume 101, number 46, 16 November 1922.

Stendahl. 1854. *La Chartreuse de Parme*. First published 1839. Paris: Adolphe Delahays.

Stendahl. 2016. *Le Rouge et Le Noir*. First published 1830. Niort: Atlantic Editions.

Stenger, Victor J. 2011. *The Fallacy Of Fine Tuning: Why The Universe Is Not Designed For Us*. Amherst: Prometheus.

Stenger, Victor J. 2012. *God and the Folly of Faith: The Incompatibility of Science and Religion*. Amherst: Prometheus.

Stenonis, Nicolai. 1910. *Prœmium Demonstrationum Anatomicarum*. First published 1673. Opera Philosophica, Volume 2, Treatise 31. Edited by Vilhelm Maar. Copenhagen: Vilhelm Tryde.

Stephens, Alexander. 1866. *The Corner Stone Speech, delivered 21 March 1861*. Alexander H. Stephens in Public and Private, with Letters and Speeches, Before, During and Since the War. Edited by Henry Cleveland. Philadelphia: National Publishing Company.

Stewart, Ian. 2008. *Taming the Infinite: The Story of Mathematics*. London: Quercus.

Stiglitz, Joseph. 2011. Of the 1%, by the1%, for the 1%. *Vanity Fair*. 31

March 2011.

Stobaeus, Joannes. 2016. *Apophthegms of the Seven Sages*. First published circa 450 BCE. Loeb Classical Library, LCL 525. Translated by André Laks and Glenn W. Most. Cambridge: Harvard University Press.

Storr, Anthony. 1991. *Human Destructiveness*. New York: Grove Weidenfeld.

Stowe, Harriet Beecher. 1896. *Uncle Tom's Cabin: Or, Life Among the Lowly*. First published 1852. Boston: Houghton Mifflin.

Strachey, John. 1933. *The Coming Struggle for Power*. Second edition. New York: Covici Friede.

Suarès, André. 1915. *Péguy*. Paris: Émile-Paul Fréres.

Sutcliffe, Thomas. 2008. Let Those Without Sin Cast the First Stone. *The Independent*. 11 March 2008.

Swift, Jonathan. 1982. *Thoughts on Religion*. First published circa 1765. The Portable Swift. Edited by Carl Van Doren. New York: Penguin.

Syrus, Publilius. 1954. *Aphorisms (Sententiae)*. First published circa 50 BCE. Loeb Classical Library LCL 284. Translated by J. Wight Duff and Arnold M. Duff. Cambridge: Harvard University Press.

Taber, Henry. 1897. *Faith or Fact*. New York: Peter Eckler.

Tacitus, Publius Cornelius. 1956a. *The Annals, Volume 2*. First published circa 110 CE. Loeb Classical Library, LCL 249. Translated by John Jackson. Cambridge: Harvard University Press.

Tacitus, Publius Cornelius. 1956b. *The Annals, Volume 3*. First published circa 110 CE. Loeb Classical Library, LCL 312. Translated by John Jackson. Cambridge: Harvard University Press.

Tegmark, Max. 2018. *What Is and What Matters*. Waking Up Podcast with Sam Harris. Episode 120, 19 March 2018.

Tennessee v John Thomas Scopes. 1925. The World's Most Famous Court Trial: Tennessee Evolution Case. University of Minnesota Law Library. Third edition. Cincinnati: National.

Tertullian. 1998. *Apology (Apologeticus)*. First published 197 CE. Loeb Classical Library, LCL 250. Translated by T. R. Glover. Cambridge: Harvard University Press.

Thatcher, Margaret. 1965. *Speech to National Union of Townswomen's Guilds Conference*. Delivered 20 May 1965. Margaret Thatcher Foundation.

Thatcher, Margaret. 1975. *Speech to Chelsea Conservative Association*. Delivered 26 July 1975. Margaret Thatcher Foundation.

Thatcher, Margaret. 1989. *Speech to Conservative Party Conference*. Delivered 13 October 1989. Margaret Thatcher Foundation.

Thatcher, Margaret. 1991. *Speech to the Industrial League of Orange County*. Delivered 14 March 1991. Margaret Thatcher Foundation.

Thiry, Paul-Henri. 1771. *Le Systèm de la Nature, Volume 2*. Published under the pseudonym Monsieur Mirabeau. London: Publisher unknown.

Thiry, Paul-Henri. 1772. *Le Bon-Sens; Ou Idées Naturelles Opposées aux Idées Surnaturelles*. London: Publisher unknown.

Thoreau, Henry David. 1993. *Civil Disobedience*. First published 1849. Civil Disobedience and Other Essays. New York: Dover Publications.

Thucydides. 2003. *The History of the Peloponnesian War*. First published circa 400 BCE. Great Books of the Western World, Volume 5. Edited by Mortimer J. Adler, translated by Richard Crawley, seventh printing. Chicago: Encyclopædia Britannica.

Tocqueville, Alexis de. 2003. *Democracy in America*. First published 1835-1840. Great Books of the Western World, Volume 44. Edited by Mortimer J. Adler, seventh printing. Chicago: Encyclopædia Britannica.

Tolstoy, Leo. 1985. *Sevastopol in May*. First published 1855. Tolstoy: Tales of Courage and Conflict. Edited by Charles Nieder. New York: Carroll & Graf.

Torcaso v Watkins. 1961. United States Supreme Court. Justice Hugo Black leading, citation 367 US 488, 19 June 1961

Treaty of Perpetual Peace and Friendship Between the United States of America and the Bey and Subjects of Tripoli of Barbary. 1797. American State Papers, Senate, Fifth Congress, First Session, Foreign Relations, Number 122. Ratified on 7 June 1797. Washington, D.C.: Library of Congress.

Trenchard, John, and Thomas Gordon. 1720. *Of Freedom of Speech: That the Fame Is Inseparable From Publick [sic] Liberty*. Cato's Letters: Or Essays on Liberty, Civil and Religious, and Other Important Subjects, Volume 1. First published 1737. Published under the pseudonym Cato, number 15, 4 February 1720.

Truman, Harry S. 1963. *The President's Special Conference With the Association of Radio News Analysts*. 13 May 1947. Public Papers of the Presidents of the United States: Harry S. Truman, 1 January to 31 December1947. Washington: United States Printing Office.

Tuchman, Barbara. 1979. *The Distant Mirror: The Calamitous Fourteenth Century*. First published 1978. New York: Ballantine.

Tuchman, Barbara. 1980. *Stilwell and the American Experience in China, 1911-45*. First published 1971. Toronto: Bantam.

Tuchman, Barbara. 2014. *The Proud Tower: A Portrait of the World Before the War, 1890-1914*. First published 1966. New York: Random House.

Turing, Alan. 1950. Computing Machinery and Intelligence. *Mind: A Quarterly Review of Psychology and Philosophy*. Volume 59, number 236, October 1950, pp. 433-460.

Twain, Mark. 1940. *Mark Twain in Eruption: Hitherto Unpublished Pages About Men and Events*. Edited by Bernard DeVoto. New York: Harper & Brothers.

Twain, Mark. 1974. *Letters From the Earth*. First published 1962. Edited by Bernard DeVoto.

Twain, Mark. 1989. *Following the Equator*. First published 1897. New York: Dover.

Tyson, Neil deGrasse. 2001. The Beginning of Science. *Natural History Magazine*. Volume 110, number 1, February 2001, pp. 80-83.

Tyson, Neil deGrasse. 2011. *Real Time with Bill Maher*. Hosted by Bill Maher, episode 201, 4 February 2011.

Tyson, Neil deGrasse. 2018. *Twitter (neiltyson)*. 14 April 2018, 11:22 a.m. USA CST.

Tyson, Neil deGrasse. 2019. *Twitter (neiltyson)*. 8 May 2019, 10:22 a.m. USA CST.

Ubbelohde, A. R. 1963. *Man and Energy*. First published 1954. Baltimore: Penguin.

Universal Declaration of Human Rights. 1948. United Nations. Adopted 10 December 1948.

Vahdat, Ahmed. 2019. Iranian Women Face 10 Years for Sharing Videos of Hijab Removal. *The Telegraph*. 29 July 2019.

Vakil, Dina. 2008. Don't Allow Religious Hooligans to Dictate Terms. *The Times of India*. 16 January 2008.

Valéry, Paul. 1943. *Tel Quel*. Eighteenth edition. Paris: Gallimard.

Veenhoven, Ruut. 2007. *Measures of Gross National Happiness*. Presented at OECD World Forum, Istanbul, 30 June 2007.

Verne, Jules. 1864. *Voyage au Centre de la Terre*. Paris: J. Hetzel.

Vinci, Leonardo da. 1883. *The Literary Works of Leonardo da Vinci, Volume 2*. Edited by Jean Paul Richter. London: Sampson, Low, Marston, Searle & Rivington.

Virginia Assembly. 1779. *A Bill for Establishing Religious Freedom*. 18 June 1779. National Archives.

Voltaire. 1785. *Idées Républicaines*. First published 1765. Politique et Legislation, Volume 1 (within Oeuvres Completes de Voltaire, Volume 34). Edited by Jean-Antoine-Nicolas de Caritat and Jacques-Joseph-Marie Decroix. Kehl: Société Littéraire Typographique.

Voltaire. 1801. *L'Évangile de la Raison*. Publication date expressed as year 10 on French Republican calendar. Publisher unknown.

Voltaire. 1816a. *Dieu, Dieux*. First published 1752. Dictionnaire Philosophique de Voltaire, Volume 6, Lettres DEJ - EGL. Paris: Didot.

Voltaire. 1816b. *Secte*. First published 1752. Dictionnaire Philosophique de Voltaire, Volume 13, Lettres QUI - T. Paris: Didot.

Voltaire. 1874. *A L'Auteur du Livre*. First published 1769. Épitres, Satires, Contes, Épigrammes de Voltaire: Suivis de Fragments de la Pucelle.

Voltaire. 1900. *Candide ou l'Optimisme*. First published 1759. Paris: Jules Tallandier.

Voltaire. 1922. *Catalogue alphabétique de la plupart des écrivains français qui ont paru dans le siècle de Louis XIV*. First published 1751. Le Siècle de Louis XIV, Volume 2. Paris: Garnier Frères.

Vonnegut, Kurt. 1999. *God Bless You, Dr. Kevorkian*. New York: Seven Stories Press.

Wallace, Alfred Russell. 1863. Remarks on the Rev. S. Haughton's Paper on the Bee's Cell, and on the Origin of Species. *Annals and Magazine of Natural History*. Series 3, volume 12, issue 70, 1863, pp. 303-309.

Warburton, Nigel. 2009. *Free Speech: A Very Short Introduction*. Oxford: University Press.

Washington, George. 1790. *George Washington to Catherine Macaulay Graham, 9 January 1790*. George Washington Papers. Series 2, Letterbooks 1754-1799: Letterbook 17, 5 May 1789 - 7 September 1791. Washington, D.C.: Library of Congress.

Washington, Booker T. 1906. *Up From Slavery: An Autobiography*. New York: Doubleday Page.

Watson, Richard. 1796. *An Apology for the Bible*. Dublin: P. Wogan.

Watterson, Bill. 1992. *Scientific Progress Goes Boink*. New York: Scholastic.

Waugh, Evelyn. 1962. *Decline and Fall*. First published 1928. London: Chapman and Hall.

Weber, Max. 1946. *Essays in Sociology*. First published 1919. Translated and edited by H. H. Gerth and C. Wright Mills. New York: Oxford University Press.

Weil, Simone. 1978. *Lectures on Philosophy*. First published 1959.

Translated by Hugh Price. Cambridge: Cambridge University Press.

Weil, Simone. 1981. *Two Moral Essays: Draft for a Statement of Human Obligations and Human Personality*. Pendle Hill Pamphlet 240. Edited by Ronald Hathaway. Wallingford: Pendle Hill.

Weinberg, Steven. 2006. *Beyond Belief: Science, Reason, Religion & Survival*. Salk Institute. The Science Network, session 1, 5-7 November 2006.

Weiss v District Board of School District Number 8 of the City of Edgerton. 1890. Supreme Court of Wisconsin. Citation 76 Wis. 177, 18 March 1890.

Wells, H. G. 1934. *The Outline of History: Being a Plain History of Life and Mankind*. First published 1920. London: Cassell.

Wells, H. G. 2000. *Crux Ansata: An Indictment of the Roman Catholic Church*. First published 1944. Escondido: Book Tree.

Wels, Otto. 1933. *23 März 1933: Rede von Otto Wells*. Speech to German parliament on 23 March 1933. Social Democratic Party of Germany (SPD). [spd.berlin]

Wesley, John. 1826. *An Extract of the Rev. John Wesley's Journal, No. XV, from May 14, 1768 to September 1, 1770*. The Works of the Rev. John Wesley in Ten Volumes, Volume 3. First American edition, journal entry from 14 May 1768. Philadelphia: Neall and Stockton.

Westermarck, Edward. 1917. *The Origin and Development of the Moral Ideas, Volume 2*. First published 1906. Second edition. London: MacMillan.

Wheeler, John Archibald. 1989. *Information, Physics, Quantum: The Search For Links*. Proceedings of the 3rd International Symposium on Foundations of Quantum Mechanics in Light of New Technology. Tokyo, pp. 354-368.

Whewell, William. 1847. *The Philosophy of the Inductive Sciences, Founded Upon Their History, Volume 2*. First published 1840. London: John W. Parker.

White, E. B. 1943. The Meaning of Democracy. *The New Yorker*. 3 July 1943.

Whitehead, Alfred North. 1927. *Religion in the Making: Lowell Lectures, 1926*. Cambridge: Cambridge University Press.

Whitehead, Alfred North. 1956. *Dialogues of Alfred North Whitehead*. Recorded by Lucien Price. New York: Mentor.

Whitman, Walt. 1882. *Specimen Days & Collect*. Philadelphia: Rees Welsh & Co.

Wiesel, Elie. 1992. Have You Learned the Most Important Lesson of All? *Parade Magazine*. 24 May 1992.

Wilde, Oscar. 1899. *An Ideal Husband*. First published 1895. London: Leonard Smithers.

Wilde, Oscar. 1905. *The Critic as Artist, Part II*. Intentions. New York: Brentano's.

Wilkins, John. 1708. *Mathematical Magick*. First published 1648. The Mathematical and Philosophical Works of the Right Reverend John Wilkins. London: J. Nicholson.

Williams, H. A. 2002. *The True Wilderness*. First published 1965. London: Continuum.

Williamson, Mark. 1992. Haldane's Special Preference. *The Linnean: Newsletter and Proceedings of the Linnean Society of London*. Edited by B. G. Gardiner, volume 8, number 3.

Willis, Ellen. 2001. Freedom From Religion. *The Nation*. 12 February 2001.

Willson, James R. 1832. *Prince Messiah's Claims to Dominion Over All Government and the Disregard of His Authority by the United States in the Federal Constitution*. Albany: Packard, Hoffman and White.

Wilson, Woodrow. 1912. *An Address to the New York Press Club*. The Papers of Woodrow Wilson, Volume 25. Edited by Arthur S. Link, speech delivered 9 September 1912.

Wilson, Woodrow. 1919. *Last Hopes of Peace: President Wilson's Address to the Senate of the United States, January 22, 1917*. President Wilson's Great Speeches and Other History Making Documents. Chicago: Stanton & Van Vliet.

Wilson, Michael. 1968. *Planet of the Apes*. First published 1967. Performance by Charlton Heston. Twentieth Century Fox. [screenplay]

Wolfe, Tom. 1980. *Entr'actes and Canapes*. In Our Time. New York: Farrar Straus Giroux.

Wollstonecraft, Mary. 1833. *Vindication of the Rights of Woman, with Strictures on Political and Moral Subjects*. First published 1792. New York: A. J. Matsell.

Wong, Joshua. 2019. *Twitter (joshuawongcf)*. 10 August 2019, 9:04am USA CST.

Wood, James. 1893. *Dictionary of Quotations*. London: Frederick Warne.

Woods, Amanda. 2019. Iranian Civil Rights Activist Gets Prison for Taking Off Hijab in Public. *The New York Post*. 29 August 2019.

Wootton, Barbara. 1950. *Testament for Social Science: An Essay in the Application of Scientific Method to Human Problems*. New York: W. W.

Norton.

The World This Week. 2019. *The Economist*. 6 July 2019.

Wright, Chauncey. 1958. *The Philosophical Writings of Chauncey Wright: Representative Selections*. Edited by Edward H. Madden. New York: Liberal Arts Press.

Wright, Richard. 1998. *Native Son*. First published 1940. New York: Perennial Classics.

Wright v Houston Independent School District. 1972. U.S. District Court for the Southern District of Texas. Woodrow Seals presiding, citation 366 F. Supp. 1208, 3 August 1972.

Wu, Chien-Shiung. 1963. Queen of Physics. *Newsweek*. Volume 61, number 20, 20 May 1963, pp. 92-95.

Youngman, Henny. 1981. *Henny Youngman's 500 All-Time Greatest One-Liners*. New York: Pinnacle.

Yousafzai , Malala. 2013. Malala Yousafzai Delivers Defiant Riposte to Taliban Militants with Speech to the UN General Assembly. *The Independent*. 12 July 2013.

Zauzmer, Julie. 2016. The Man Billed as the First Atheist Who Could Be Bound For Congress Isn't Actually an Atheist At All. *The Washington Post*. 4 May 2016.

The Zend-Avesta. 1880. *The Zend-Avesta, Part 1, The Vendidad*. The Sacred Books of the East, Volume 4. Translated by James Darmesteter, edited by F. Max Müller. Oxford: Clarendon Press

Zinn, Howard. 1990. *Declarations of Independence: Cross-Examining American Ideology*. New York: HarperCollins.

Zinn, Howard. 1999. *A People's History of the United States: 1492-Present*. First published 1980. Twentieth Anniversary Edition. New York: HarperCollins.

Zinn, Howard. 2008. *War and Social Justice*. Transcript of speech delivered at Binghamton University in November 2008.

Zola, Émile. 1898. J'Accuse: Lettre au Président de la République. *L'Aurore*. Second year, number 87, 13 January 1898.

Zorach v Clausen. 1952. United States Supreme Court. Justice Robert Jackson dissenting, citation 343 US 306, 28 February 1952.

De omnibus dubitandum est

ORBIS
UNUM
PRESS

An imprint of La Jolie Vie, LLC